Fluorescent Chemosensors for Ion and Molecule Recognition

ACS SYMPOSIUM SERIES **538**

Fluorescent Chemosensors for Ion and Molecule Recognition

Anthony W. Czarnik, EDITOR

Ohio State University

Developed from a symposium sponsored
by the Division of Organic Chemistry
at the 204th National Meeting
of the American Chemical Society,
Washington, DC,
August 23–28, 1992

American Chemical Society, Washington, DC 1992

Library of Congress Cataloging-in-Publication Data

Fluorescent chemosensors for ion and molecule recognition /
Anthony W. Czarnik, editor.

p.   cm.—(ACS symposium series, ISSN 0097–6156; 538)

"Developed from a symposium sponsored by the Division of Organic
Chemistry of the American Chemical Society at the 204th Meeting of
the American Chemical Society, Washington, DC, August 23–28, 1992."

Includes bibliographical references and index.

ISBN 0–8412–2728–4

1. Biosensors—Congresses. 2. Fluorescence—Congresses.

I. Czarnik, Anthony W. II. American Chemical Society. Division of
Organic Chemistry. III. American Chemical Society. Meeting (204th:
1992: Washington, D.C.) IV. Series.

R857.B54F58   1993
681′.761—dc20

93–21067
CIP

Foreword

THE ACS SYMPOSIUM SERIES was first published in 1974 to provide a mechanism for publishing symposia quickly in book form. The purpose of this series is to publish comprehensive books developed from symposia, which are usually "snapshots in time" of the current research being done on a topic, plus some review material on the topic. For this reason, it is necessary that the papers be published as quickly as possible.

Before a symposium-based book is put under contract, the proposed table of contents is reviewed for appropriateness to the topic and for comprehensiveness of the collection. Some papers are excluded at this point, and others are added to round out the scope of the volume. In addition, a draft of each paper is peer-reviewed prior to final acceptance or rejection. This anonymous review process is supervised by the organizer(s) of the symposium, who become the editor(s) of the book. The authors then revise their papers according to the recommendations of both the reviewers and the editors, prepare camera-ready copy, and submit the final papers to the editors, who check that all necessary revisions have been made.

As a rule, only original research papers and original review papers are included in the volumes. Verbatim reproductions of previously published papers are not accepted.

M. Joan Comstock
Series Editor

Contents

INDEXES

Preface

IN THE FLUORESCENT CHEMOSENSOR FIELD, three conditions have impressed me in the past five years. First, this area represents an unusual opportunity for the organic chemistry community to make contributions to an important technological problem. Applications of real-time sensing methods are tremendously varied. Until the invention of fluorescent chemosensors for calcium, little research was done on intracellular calcium concentrations. Once a tool became available that could be used by cell biologists as a "black box", it was rapidly assimilated into the biochemistry community; calcium sensing is now virtually a new field unto itself. There remain a great many biological analytes worth seeing in real time, among them inorganic phosphate, magnesium, sodium, potassium, and glucose. A new functional fluorescent chemosensor for any one of these species would open up an entire new field of investigation.

The same holds true for remote sensing applications. Imagine a day when blood concentrations of important analytes are determined not in a batch mode, but rather through the use of a fiber-optic bundle inserted into an artery. Each fiber is equipped with a different chemosensor, such that the real-time concentrations of several dozen analytes are provided on a monitor at the patient's bedside. Devices for real-time monitoring of blood pH, pO_2, and pCO_2 that are now available use fluorescent chemosensors. Technical issues, some of them daunting, await new blood monitoring devices; nevertheless, the payoff seems worth the research investments.

The second fact, somewhat unsettling, is that the communities that need chemosensors and those that might synthesize them are largely unaware of each other. Application-driven communities such as cell biologists, environmental scientists, electrical and electronic engineers, medical diagnosticians, and researchers in chemical warfare agent detection are virtually unaware that a branch of chemistry exists whose goal is to design and synthesize receptors for ions and molecules. Most of these communities are aware of progress in biosensor research because useful selectivities already exist in biotic receptors such as antibodies. However, the argument can readily be made that chemosensors represent an important complement to biosensors. Compared to chemosensors, biosensors are likely to be expensive and unstable. Antibody binding tends to be irreversible and yields dosimeters rather than sensors. Biosensors based on enzyme catalysis are reagent based and therefore of limited lifetime by design. Finally, because high-affinity antibodies cannot be raised easily against small molecules or ions, biosensors for abiotic species (e.g., lead ion) are difficult to envision.

The third puzzling fact is that funding organizations have shown interest in the production of working devices but virtually no interest in supporting the more basic research that is needed to learn how to make them. For example, calcium chemosensors are almost all based on chelation by the EGTA ligand, whose affinity for Ca(II) has been documented for many years. Agencies have shown little tolerance for proposed research acknowledging that sensors for other analytes will not immediately demonstrate the same quality of binding. Thus, the chemist seeking to do chemosensor research is placed in the position of having to propose the construction of new sensors with properties that likely cannot be achieved in the time frame of a grant cycle. Overcoming this obstacle will require as much creativity as that needed to solve the research issues.

The purpose of this book is to provide a starting point that is greater than zero so that both chemists and potential end-users can begin the adventure that lies ahead.

ANTHONY W. CZARNIK
Current address:
Parke–Davis Pharmaceutical Research
2800 Plymouth Road
Ann Arbor, MI 48105

June 14, 1993

Chapter 1

Supramolecular Chemistry, Fluorescence, and Sensing

Anthony W. Czarnik[1]

Department of Chemistry, Ohio State University, Columbus, OH 43210

Each new ability to visualize analyte concentrations in real-time portends the fabrication of new sensors. Potential applications include fiber optic-based remoting sensing of solution species and the monitoring of concentrations within living cells. A fluorescent chemosensor is a compound of abiotic origin that complexes to an analyte reversibly with concomitant fluorescence signal transduction. Such chemosensors have been used for fluorimetric metal ion determination for over 100 years. The advent of ligand engineering has introduced a more systematic approach to the design of chemosensors with new selectivities and signal transduction schemes. The technological driving force to achieve useful chemosensors will stimulate investigation of new topics in molecule recognition, fluorescent signal transduction, and their intersection.

A *sensor* is a device that interacts with matter or energy and yields a measurable signal in response. Many scientists to date have considered sensors as macroscopic devices, such as the pH electrode (an H^+ sensor) or the bimetallic strip (a temperature sensor). However, with the increasing recognition of nanotechnology there remains little reason not to view suitably engineered individual molecules as devices capable of performing useful work on other molecules. A catalyst is therefore a construction device able to fabricate new compounds. A transport protein is a transportation device able to move species through barriers.

---

[1]Current address: Parke–Davis Parmaceutical Research, 2800 Plymouth Road, Ann Arbor, MI 48105

NOTE: Portions of this chapter have been adapted from the Ph.D. theses of Michael Huston (1990) and Scott Van Arman (1990), Ohio State University.

Likewise, phenolphthalein is a sensing device able to signal the concentration of hydrogen ion. As such, phenolphthalein is an abiotic molecule that senses an abiotic analyte. Because living organinisms have provided such a rich array of both analytes worth sensing and receptors for many of them, biotic sensors have received considerable attention. As we begin a discussion of fluorescent chemosensors, a definition of terms is in order. *Chemical sensor*: a micro- or macroscopic device that interacts reversibly with a chemical analyte with signal transduction. *Chemosensor*: a molecule of abiotic origin that signals the presence of matter or energy. *Biological sensor*: a micro- or macroscopic device that interacts reversibly with a biological analyte with signal transduction. *Biosensor*: a molecule of biotic origin that signals the presence of matter or energy. *Intrinsic fluorescence probe*: a chemosensor in which the mechanism for signal transduction involves interaction of the analyte with a ligand that is part of the fluorophore π-system. *Conjugate fluorescence probe*: a chemosensor in which the mechanism for signal transduction involves interaction of the analyte with a ligand electronically-insulated from the fluorophore π-system.

Thus, phenolphthalein is a chemosensor. In the presence of a sufficient concentration of OH⁻, phenolphthalein is deprotonated with visible light signalling at 552 nm. *Signal transduction* is the mechanism by which an interaction of sensor with analyte yields a measurable form of energy, which may reside in wavelength regimes characterized by various spectroscopies (e.g., UV, visible, NMR) or may yield electrochemical responses. In fact, each mode of signal transduction occupies an entire literature whose summary is beyond the intent of this chapter. Rather, the function of this symposium was to survey contemporary research into the synthesis of abiotic molecules that sense analyte concentrations with fluorescent signal transduction: *fluorescent chemosensors*.

Statement Of The Research Topic

The requisite research issues essential to the creation of fluorescent chemosensors are: (1) *how can one bind a molecular entity with selectivity (preferably from water)*, (2) *what molecular changes result in fluorescence changes*, and (3) *what mechanisms for binding and fluorescence signal transduction intersect.*

Binding a Molecular Entity with Selectivity. While much of organic chemistry focusses on the selective creation of covalent bonds, such irreversible associations of potential analytes yield chemodosimeters (designed for cumulative assay) in contrast to chemosensors (designed for real-time assay). Thus, sensing applications require the existence of receptors (hosts) that associate with analytes (guests) selectively *and reversibly*.

One approach to the fabrication of such hosts is to utilize the many such receptors available from biological sources. Selective binding of species is demonstrated by enzymes for their substrates (and cofactors), by transport proteins for their transportees, and by antibodies for their antigens. In particular, because antibodies can be raised to many antigens, the biotic creation of hosts proves a powerful method for binding analytes with selectivity. Indeed, it is fair to say that biotic receptors have realized far greater practical utility to date than have synthetic, abiotic receptors. Biosensors are sensors derived from such biotic receptors.

Having said this, there is a strong case to be made for the complementary, abiotic approach to receptor formulation. Except for antibodies, the analytes amenable to biotic receptor fabrication are limited to those provided by nature; there are many analytes of interest, including small molecules and metal ions, against which high affinity antibodies cannot be raised. For example, it is not at present possible to raise a high affinity antibody against glucose. There is a sizable class of analytes for which the biotic approach to selctive binding is not likely to be successful even in principle. Additionally, biotic receptors often have limited stabilities that make their translation from the benchtop to the field difficult. And sometimes they are not readily amenable to scaled-up production, which makes them expensive.

An alternative approach to the fabrication of selective receptors for small molecules is to design and synthesize one from scratch. This field of chemistry has been in existance for perhaps 30 years, and the subfield has been referred to as *supramolecular chemistry (1)*. Lessons learned from inspection of biotic host-guest interactions provide some guidance for the engineering of synthetic receptors. The non-peptidic frameworks used in the synthesis of abiotic receptors typically engender enhanced stability. However, the efforts required to chemically synthesize molecules with either cleft-like or cavity-like binding sites is often substantial and rate-limiting to their study. Additionally, not all of the synthetic receptors evaluated to date are water soluble or compete effectively with water for guest complexation. However, the abiotic approach to receptor fabrication provides at present the only intellectually-viable approach to the design of chemosensors for many analytes of interest. While synthetic schemes may be long and some steps low yielding, the synthesis of even a few milligrams of a fluorescent chemosensor might well be sufficient for thousands of analyses.

Molecular Changes That Result in Fluorescence Changes.

A Brief Introduction to Fluorescence. In 1565 the Spanish physician and botanist Nicolas Monardes made the first recorded observation of fluorescence when he noted a strange blue glimmer from water contained in a cup made from a specific wood (*Ligirium nephiticiem*). In the 17'th century this phenomenon was more extensively described by Robert Boyle and Isaac Newton. It was not until 1845 that the first crude fluorescence emission spectrum was obtained for quinine by John Herschel. Great strides were made toward the understanding of fluorescence in the late 19'th century by George Stokes. He established the technique of observing fluorescence with two different colored filters, one for the excitation beam, and one through which to observe the emission. In 1852 Stokes became the first to determine and report that emission was at a longer wavelength than the excitation. This characteristic of fluorescence is now known as Stokes' Law. Stokes also described the relationship between fluorescence intensity and concentration. He outlined the quenching of fluorescence at high concentration and by the presence of foreign substances and used this information to propose the use of fluorescence for the detection of organic substances. In 1867 F. Goppelsroder developed a method for the determination of non-fluorescent Al(III) by forming a strongly fluorescent morin chelate, which was the first analysis based on fluorescence.

The contemporary explanation for the fluorescence phenomenon may be summarized as follows. The absorption of a quantum of light by a molecule results in the elevation of an electron from the molecule's ground electronic state (S_o) to one of several vibrational levels in the an electronic excited state. In solution, the excited state molecule rapidly relaxes to the lowest vibrational level of the lowest electronic state (S_1). The energy thus stored in the excited state may be released in several ways. The electron may return to the electronic ground state with only the release of heat (radiationless relaxation). After relaxing thermally to the lowest vibrational level of the S_1 state, the electron may return to the S_0 state with light emission (fluorescence). Or, if the molecule is sufficiently long-lived in the S_1 state, it may cross into a lower energy triplet state (T_1). Relaxation from the T_1 state to the S_0 state can also occur with light emission in solids (phosphorescence), by the release of energy (radiationless transition), or by chemical reaction.

Fluorescence as a Signal Transduction Mechanism. There are many reasons for which fluorescence might be identified as the optimal signal transduction mechanism in potential sensing applications. Fluorescence is an enormously sensitive technique, in large part because the observing wavelength is always longer than that of the exciting wavelength. Thus, given appropriate choice of fluorophores, a signal may be read versus zero or near-zero background. In the extreme, the fluorescence of even single molecules has been observed. In some applications (e.g., water quality monitoring), background fluorescence is not a particular problem. However, the autofluorescence of most biological samples generates background signal up to about 650 nm. Long wavelength fluorophores (i.e., near-IR) generally avoid these interferences, although sometimes at the cost of severely narrowing the signal transduction mechanisms that function in such compounds. Fluorescence signalling also permits the monitoring of both excitation and emission wavelengths. The emission signal may be observed in the form of intensity, intensity ratio, or lifetime measurements. We posess a rudimentary ability to engineer molecules with predictable responses to the first two signal types, if not yet the third. Fluorescence is usually nondestructive.

Electrochemical methods benefit from Nernstian behavior permitting the signalling of analytes over very large concentration ranges (e.g., pH electrodes), while spectroscopic signalling is in general limited in range to perhaps two concentration decades. This has the effect of requiring that the K_D for a given fluorescent chemosensor be engineered near to the actual concentration of the analyte of interest. However, light can travel through and egress from environments without an absolute requirement for a physical waveguide; electrochemical sensors rely on wires. This proves enormously powerful in, for example, the simultaneous visualization of concentrations in all regions of a living cell. Waveguides *can* be used advantageously, however, in conjunction with fluorescence signalling. While the laser, fiber optic, and detector technologies are quite well established to permit a vast array of remote sensing applications, *what is lacking is the scientific basis upon which to design and synthesize fluorescent chemosensors with appropriate selectivities that translate to immobilization on the tips of fiber optics.*

In listing the criteria required for fluorescent chemosensors that would yield functional, real-time fiber optic sensors, we delineate the broad scientific questions to be addressed.

(a) *The binding domain must have sufficient selectivity for the analyte of interest as compared to others present.*

(b) *The analyte must be capable of dissociating from the binding domain in real-time.*

(c) *There must exist a signal transduction mechanism between the binding and the fluorescing domains.*

(d) *The ideal system will incorporate a method for internal calibration.*

(e) *The excitation wavelength should correspond to that of an inexpensive, portable laser source.*

(f) *The emission wavelength should be consistent with inexpensive detectors and removed from that of adventitious fluorescent impurities in the media to be monitored.*

(g) *The conjugate fluoroionophore (i.e., both binding and fluorescing domains) should prove stable to both hydrolysis and oxidation reactions.*

(h) *The fluorescence signal being measured should not vary with pH in the range encountered in the environment.*

(i) *The signal being measured should not be substantially modulated by adventitious fluorescence quenchers encountered in the environment.*

(j) *The above properties of the chemosensor should remain largely intact upon immobilization onto the waveguide material.*

Even in doing so, it must be acknowledged that no chemosensors are yet known that meet all these criteria. Fluorescence is the ideal signal transduction mechanism for fiber optic remote sensing applications. In the near future, the applications that await chemosensor synthesis will serve as a major driving force for basic research into the organic chemistry of fluorescent chemosensors.

Chemical Mechanisms for Fluorescence Modulation. While UV/visible signalling almost always results from the ionization of a conjugated substituent, there exists a plethora of mechanisms by which fluorescence signal transduction may be engendered. It is useful to categorize the mechanisms for fluorescence modulation described to date *via* the type of measurement that is made. These are: *intensity, intensity-ratio,* and *lifetime.* A pictoral summary of each is found in Figure 1 of the chapter by Szmacinski and Lakowicz.

Intensity measurement. By far the most common mode of fluorescence modulation described to date is the increase or decrease of fluorescence intensity at a single emission wavelength upon analyte binding. Methods for halide ion and molecular oxygen analyses rely on non-chelative (collisional) quenching by these species. There is a substantial literature describing chelative metal ion detection in water using intrinsic fluorescent chemosensors, in which donor atoms of the ligand are a part of the fluorophore's structure. The complexation of a metal ion results in either enhanced fluorescence of the chemosensor ('chelation-enhanced fluorescence', or **CHEF**), or in decreased fluorescence ('chelation-enhanced quenching', or **CHEQ**). In such intrinsic fluorescent chemosensors, **CHEF** typically results upon chelation of one of the ions that is not inherently quenching (non-redox active, closed-shell, e.g., Zn(II), Cd(II), Al(III)) and **CHEQ** is observed on chelation of an inherently quenching ion (e.g., Cu(II), Hg(II), Ni(II)). The chapter by Bell describes the use of intrinsic sensors with intensity signalling. It is this author's conclusion that there is no single mechanistic rationale that can be used to explain **CHEF** in intrinsic fluorescent chemosensors.

In 1977, the application of supramolecular chemistry to fluorescent chemosensor design took root with the publication of an article by Sousa describing naphthalene-crown ether compounds that bind alkali-metal ions with changes in emission intensity (either **CHEF** or **CHEQ**, depending on the chemosensor) (2). Because no donor atom was a part of or conjugated to the fluorophore π-system, these are examples of conjugate fluorescent chemosensors. The observed changes were attributed to a heavy atom effect (for Cs^+ and Rb^+) and/or a complexation induced change in triplet energy relative to ground and excited singlet state energies. The chapter by Sousa describes other examples of intensity changes arising *via* these mechanisms.

Interestingly, a chemosensor for calcium that fits the definition of a conjugate fluorescent chemosensor was reported in 1956 by Diehl and Ellingboe (3). *Calcein* can be viewed as two iminodiacetate ligands conjugated to fluorescein by methylene spacers. Basic solutions of calcein undergo a color change from brown to yellow-green upon addition of calcium ions. The development of calcein is best considered an isolated incident in that the fluorescent properties of the molecule and its metal complexes were not studied in detail (in fact, the authors caution against carrying out titrations of calcium in the presence of strong or fluorescent light).

Intensity modulations also result when photoinduced electron transfer (**PET**) from a donor atom in a conjuage sensor is enhanced or inhibited. For example, triethylamine quenches the fluorescence of anthracene, with greater quenching efficiency at higher concentrations. When the amine is an integral, but nonconjugated, part of the fluorophore structure its effective concentration can greatly exceed an actual concentration achievable in solution. Benzylic amines appear uniquely effective at efficient quenching of anthracene (but not of many longer wavelength fluorophores). Thus, intramolecular **PET** can be very efficient in such systems. Protonation or metal ion complexation of the amine effectively . prevents the **PET** mechanism, resulting most often in **CHEF**. The chapters by de Silva, Czarnik, and Masilamani elaborate on sensing systems utilizing this mechanism for signal transduction. (If fluoroionophores undergoing photoinduced intramolecular charge transfer exhibit a shift of their emission and/or excitation spectra upon complexation, intensity-ratio measurements are possible.)

Fluorescence quantum yield (intensity) can be strongly influenced by the polarity of the microenvironment. With creative molecule engineering, the microenvironment in which a fluorophore resides can be linked physically to the presence of absence of a binding analyte. The chapter by Ueno provides examples.

Intensity-ratio measurements. Despite the desirable flexibility in the design of intensity sensors, there are drawbacks inherent to intensity-based signalling schemes. The most important drawback is that fluorescence intensity can also vary in complex samples for reasons other than analyte concentration, among them: pH changes, photobleaching (decomposition of the fluorophore mediated by the light source), compartmentalization within cells that changes microenvironment, and light scattering. This proves a particular problem in fiber optic remote sensing applications, where constant change in the chemosensor concentration requires frequent (and inconvenient) recalibration.

One approach that ameliorates many of these problems is to take intensity measurements at two excitation or emission wavelengths at which intensity responds differently; for example, an emission band at 500 nm might decrease at the same time a different band at 400 nm is increasing. Once calibrated, the ratio of intensities at these two wavelengths yields analyte concentrations independent of the absolute concentration of the chemosensor (assuming the analyte is present in large excess). Mechanistically, intensity-ratio measurements can be made using chemosensors in which both bound and unbound states are fluorescent, but with different emission maxima because a donor atom becomes deconjugated from the fluorophore π-system upon analyte chelation. Mechanisms leading to ratiometric measurements of calcium and sodium ions have been pioneered by Tsien, and examples are found in the chapters by Tsien and by Kuhn.

Nonradiative relaxation from S_1 states can be effected via fluorescence resonance energy transfer. The rate of energy transfer in donor-acceptor pairs is directly related to their distance, which can of course be affected by analyte complexation. Excitation energy transfer leads to quenching of the donor, but also to enhancement of the acceptor fluorescence; therefore, the ratio of fluorescence intensities at donor and acceptor emission wavelengths provides information on complexation via changes in donor-acceptor distance. The chapters by Valeur, Bouas-Laurent, and Krafft describe uses of the energy transfer mechanism.

A less well explored mechanism leading to ratiometry is that of ligand-induced excimer formation or dissociation between chemosensor containing two fluorophores. For example, anthracene forms an intermolecular excimer (an excited state dimer) at sufficiently high concentrations; excimer formation has the net effect of yielding ratiometric signalling (see Figure 1 of the chapter by Bouas-Laurent). Chemosensors posessing two fluorophores in their structures can form intramolecular excimers even at very low concentration, whose formation may be influenced by the binding of an analyte. The chapter by Bouas-Laurent describes examples of guest-influenced excimer formation. Also, Tsien's chapter discusses dissociation of two fluorophores induced by cyclic AMP.

One might mention also that the covalent interaction between a guest and the fluorophore substructure might also give rise to a new fluorescing species; if the new species displayed an emission maximum different from that of the unbound chemosensor, ratiometric signalling will result. One example of such a mechanism is found in the chapter by Czarnik, relating to chelatoselective Cd(II) indication.

Lifetime measurement. An alternative and newer approach to overcoming some of the shortcomings of intensity measurements is the use of signalling *via* fluorescent lifetime measurements. The accumulation of lifetime data requires more sophisticated equipment than that for intensity or intensity-ratio data, but there appear to be advantages. The measured lifetime is independent of the chemosensor concentration, which as previously described is of value. Furthermore, changes that affect a fluorophore's emission intensity often also affect its fluorescence lifetime; thus, known intensity sensors may find new utilities as lifetime sensors. The chapter by Szmacinski and Lakowicz describes their studies into the use of lifetime modulations to signal guest binding.

What Mechanisms for Binding and Signal Transduction Intersect? The collection of successful intersection strategies was the primary goal of our symposium. The invention of solutions requires an unusual degree of creativity from chemists willing to place one foot firmly in each of two research arenas. Such solutions comprise the body of this collection of contributions by researchers in the fluorescent chemosensor field.

References and Notes

(1) A.k.a. *host-guest chemistry, biomimetic chemistry,* or *molecular recognition.*

(2) Sousa, L. R.; Larson, J. M. *J. Am. Chem. Soc.* **1977**, *99*, 307.

(3) Diehl, H.; Ellingboe, J. L. *Anyl. Chem.* **1956**, *28*, 882.

A BIBLIOGRAPHY OF REVIEWS RELEVANT TO THESE TOPICS

Ion and Molecule Recognition:

Bender, M.. L.; Komiyama, M. *Cyclodextrin Chemistry,* Springer-Verlag: New York, 1977

Gutsche, C. D. *Calixarenes,* Royal Society of Chemistry: Cambridge, 1989

Gokel, G. W. *Crown Ethers & Cryptands,* Royal Society of Chemistry: Cambridge, 1991

Diederich, F. *Cyclophanes,* Royal Society of Chemistry: Cambridge, 1991

Vogtle, F. *Supramolecular Chemistry,* VCH, Weinhein, FRG, 1991

Fluorescence:

Guilbault, G. G. *Practical Fluorescence*, Marcel Dekker, Inc.: New York, 1973

Guilbault, G. G. *Practical Fluorescence*, Marcel Dekker, Inc.: New York, 1990 (this second edition does not contain all the references of the first)

Krasovitskii, B. M.; Bolotin, B. M. *Organic Luminescent Materials*, VCH: Weinheim, FRG, 1988

Lakowicz, J. R. *Principles of Fluorescence Spectroscopy*, Plenum: New York, 1983

Biophysical and Biochemical Aspects of Fluorescence Spectroscopy, Dewey, T. G., Ed., Plenum, 1991

Zander, M. *Fluorimetrie*, Springer, 1981

Reviews of Fluorescent Chemosensor Research:

Schwarzenbach, G.; Flaschka, H. *Complexometric Titrations*, (translation by H. Irving), Metheun: London, 1969

West, T. S. *Complexometry with EDTA and Related Reagents*, Broglia Press Ltd.: Bournemouth, England, 1969

Indicators, Bishop, E., ed., Pergamon: New York, 1972

Fernandez-Gutierrez, A.; Munoz de la Pena, A., in *Molecular Luminescence Spectroscopy: Methods and Applications. Part I*, Schulman, S. G., Ed., Wiley, New York: 1985, p 371.

Fiber Optic Chemical Sensors and Biosensors, Wolfbeis, O. S., Ed., CRC Press: Boca Raton, Florida, 1991, Volumes I and II

Bissell, R. A.; de Silva, A. P.; Gunaratne, H. Q. N.; Lynch, P. L.; Maguire, G. E. M.; Sandanayake, K. R. A. S. *Chem. Soc. Rev.* **1992**, *21*, 187.

Tsien, R. Y. Methods in Cell Biology 1989, 30, 127

Tsien, R. Y.; Waggoner, A. "Fluorophores for Confocal Microscopy: Photophysics and Photochemistry", in *Handbook for Biological Confocal Microscopy*, Pawley, J. B., Ed., Plenum, 1990

RECEIVED July 20, 1993

Chapter 2

Synthesis and Study of Crown Ethers with Alkali-Metal-Enhanced Fluorescence

Quest for Flashy Crowns

L. R. Sousa, B. Son, T. E. Trehearne, R. W. Stevenson, S. J. Ganion, B. E. Beeson, S. Barnell, T. E. Mabry, M. Yao, C. Chakrabarty, P. L. Bock, C. C. Yoder, and S. Pope

Department of Chemistry, Ball State University, Muncie, IN 47306

Several crown ether compounds designed to signal the presence of alkali metal cations by showing enhanced fluorescence intensity have been synthesized and studied. Examples illustrating different conceptual bases for cation-enhanced fluorescence are presented. One scheme involves the displacement by a potassium ion of a quencher complexed by a fluorescent crown ether. Another approach employs complexation by alkali metal cations to interrupt intramolecular quenching of a crown ether containing both quenching and fluorescent moieties. Metal ion enhanced fluorescence of a bis(crown ether) extraction agent is also described.

In 1977 and 1978 as part of the study of photoexcited state response to geometrically oriented perturbers, my students and I reported the observation of alkali metal ion enhanced fluorescence of a crown ether (1,8-naphtho-21-crown-6) containing an attached naphthalene chromophore (1,2). The approximately 60 percent increase in fluorescence intensity, which was observed in an ethanol glass at low temperatures, correlated with a cation induced decrease in the triplet energy level relative to the fluorescent singlet state and the ground state. The increase in fluorescence was presumably due to a decrease in the intersystem crossing rate relative to the fluorescence rate.

In 1981 my students and I began to design and synthesize compounds that would display cation enhanced fluorescence. We have focused on chromophore bearing crown ether compounds that might have fluorescence sensitive to potassium ions. We have concentrated on illustrating possible modes of action rather than developing specific analytical methods. The research process involves the following steps: 1) developing a concept or scheme, 2) envisioning a specific compound that is likely to exemplify the concept, 3) synthesizing the compound in the laboratory, 4)

0097–6156/93/0538–0010$06.00/0

studying the effect of potassium and other ions on the fluorescence intensity, and 5) evaluating the concept, and if appropriate, attempting to design more sensitive and/or more selective compounds (returning to step 3 or 1).

Several concepts (categories of action) that could result in cation-enhanced fluorescence can be envisioned. Five examples are summarized here: (a) the ion could cause subtle change(s) in energy levels or electron densities that lead to enhanced fluorescence, (b) the cation of interest could displace a quencher complexed by the crown, (c) the complexation of a cation could interrupt a quenching mechanism operable in the free crown, (d) complexation could adjust the conformation so that a new fluorescent excited state might form. (e) a crown ether used in an extraction method could promote the solubility of a fluorescent ion in a phase that is monitored for fluorescence. The literature of crown ethers contains examples employing each of these concepts (1-21) with the possible exception of case (b). Several concepts we have attempted to employ follow.

Cation Displacement of a Complexed Quencher of Fluorescence

The first concept to be considered is cation displacement of a complexed quencher of fluorescence. Scheme I illustrates the concept. To be operable the scheme requires a quencher that is complexed by a crown ether , a metal ion of interest that is not a quencher but is complexed effectively, and a crown ether ring that orients the complexed quencher so it will effectively quench the chromophore fluorescence. Scheme II shows a relatively simple crown ether system that we hoped would fulfill these requirements(22,23). The 1,5-naphtho-22-crown-6 compound was selected because of the ability of its crown ether band to hold a quencher against the face of the pi system of the naphthalene chromophore. The heavy atom ion $Cs+$ was selected as a quencher based on its propensity to increase inter-system crossing from the fluorescent S_1 state to the nonfluorescent T_1 state(1,2). It was likely, based on previous results(2), that potassium ion would be complexed by the crown, but not quench naphthalene fluorescence appreciably.

The system worked as envisioned as shown by Figures 1 and 2. Figure 1 shows the results of a fluorescence "titration", monitored at 335 nm, in methanol in which added cesium ion decreases fluorescence. Analysis of the results using a non linear least squares (NLLSQ) fitting program(24) with a quadratic equation gave a complexation constant of 4.0×10^2 M^{-1} and showed that fluorescence from a crown complexing a cesium ion is completely quenched within experimental error.

Figure 2 shows the effect of potassium ion (added as potassium acetate) when added to a methanol solution of 2.00×10^{-5} M crown with 4.00×10^{-3} M CsCl. This Figure could be considered a calibration curve for the measurement of potassium ion concentrations. An 80 percent increase in fluorescence is observed over the 0 to 0.1 M K^+ concentration range. A 2.5×10^{-3} M K^+ solution causes an 8 percent increase in fluorescence.

These data were fit by an NLLSQ program to a complicated cubic equation based on the assumption that fluorescence intensity was contributed by three species as represented by Equation 1. The species are free crown (mole fraction = X_{Cr}, fluorescence intensity = I_{fo}), crown complexed to K^+ (mole fraction = X_{CrK}, fluorescence intensity = I_{fK}), and a possible contribution from crown complexed to Cs^+

Cation Displacement of a Complexed Fluorescence Quencher
Chromophore
Quencher
Q⁻
M⁺
M⁺
Q
WEAK FLUORESCENCE DUE TO QUENCHING BY Q
STRONGER FLUORESCENCE IF M+ IS NOT A QUENCHER

Scheme I.

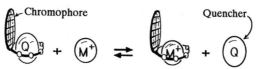

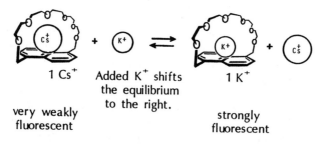

Cs⁺
K+
K+
Cs⁺
1 Cs⁺
Added K⁺ shifts the equilibrium to the right.
1 K⁺
very weakly fluorescent
strongly fluorescent

Scheme II.

CESIUM ION QUENCHING OF CROWN FLUORESCENCE

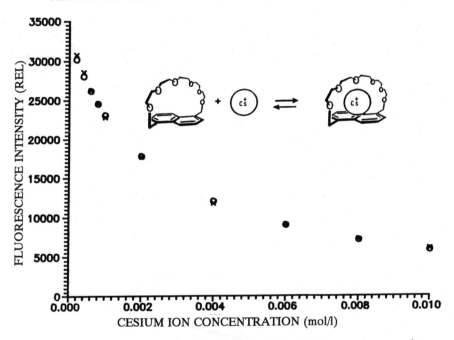

Figure 1. Relative fluorescence intensity (at 335 nm) of crown **1** versus the concentration of added CsCl. Experimental values (x) and values calculated using NLLSQ (o) are shown. As defined in the text, $[Cr]_o = 1.92 \times 10^{-5}$ molar, I_o, the initial fluorescence intensity, = 3267 (relative units); and as determined by NLLSQ, $K_{Cs} = 4.0 \times 10^2 \ M^{-1}$ and $I_{fCs} = 0$.

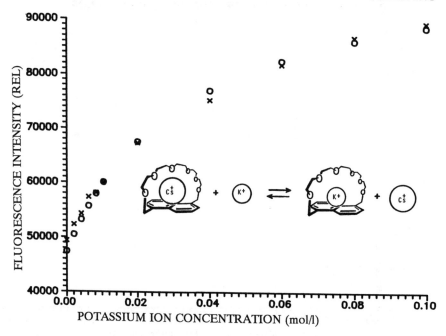

POTASSIUM ION INTERRUPTION OF CESIUM ION QUENCHING

Figure 2. Relative fluorescence intensity (at 335 nm) of crown **1** with CsCl in methanol versus the concentration of added KOAc. Experimental values (x) and NLLSQ calculated values (o) are shown. As defined in the text, $[Cr]_o$ = 2.00 x 10^{-5} M, $[Cs]_o$ = 4.00 x 10^{-3} M, I_o = 1.229 x 10^5, K_{Cs} = 4.00 x 10^2 M^{-1}, I_{fCs} = 0; and as determined by NLLSQ treatment, K = 75 M^{-1}, I_{fK} = 1.028 x 10^5 (83% of I_o).

(mole fraction = X_{CrCs}, fluorescence intensity = I_{fCs}). The NLLSQ results show that

$$I = X_{Cr}I_{fo} + X_{CrK}I_{fK} + X_{CrCs}I_{fCs} \qquad \text{(Equation 1)}$$

the crown-potassium complexation constant is 6×10^1 M^{-1} and that the complexation of potassium ion decreases fluorescence only slightly, to 85 percent of the value observed with uncomplexed crown.

The simple crown ether system of Scheme II works as envisioned; however, for some purposes it would be desireable to increase the sensitivity. It is probable that the sensitivity could be increased by synthesizing crown ethers with larger complexing constants, K_Q and K_M. What values of K_Q and K_M should one aim for to achieve sensitivity in a given range of concentrations of M^+? To answer this question we applied the following sequence of steps: 1) assume a set of reasonable values for I_o, I_{fM}, $[Q]_o$ and $[Cr]_o$; 2) use an equation that gives I (fluorescence intensity) as a function of K_Q, K_M, and $[M^+]_o$ as well as I_o, I_{fQ}, I_{fM}, $[Q]_o$ and $[Cr]_o$; 3) perform calculations of fluorescence intensity for selected values of K_Q and K_M over a range of M^+ concentrations; 4) plot the fluorescence "titration" curves that result; 5) for each K_Q K_M set, determine the concentration of M^+ that causes a 10 percent enhancement in fluorescence; and 6) for each K_Q value, plot the log of the concentration of M^+ that causes a 10 percent increase in fluorescence as a function of the log of K_M.

Figure 3 shows the calculated relationship between sensitivity and K_M at selected values of K_Q (the plot generated by step 6 above). Consider the effect of K_Q on the sensitivity. The sensitivity per log K_M increases as K_Q is increased from 10^2 to 10^4 as shown by the shift of the curves to the left. For these values of K_Q the sensitivity is most strongly influenced by background fluorescence (in the absence of M^+), and the increased complexation of quencher (with increased K_Q values) decreases that background. From $K_Q = 10^4$ on to higher values, the curves move to the right showing that K_M values must be increased almost proportionately with K_Q even to maintain the same sensitivity. For these higher values of K_Q, sensitivity is limited by the propensity of M^+ to replace the quencher Q, so higher K_M values are required. Another feature of Figure 3 is the turn toward the horizontal observed for the $K_Q = 10^2$ and $K_Q = 10^3$ curves (and presumably in curves for larger K_Q values at lower M^+ concentrations). The curves become horizontal when even complete complexation of the small amounts of M^+ present will not cause a significant fluorescence increase relative to the fluorescence background in the absence of M^+. As K_Q is made larger, the curves become horizontal at lower concentrations because a larger fraction of crown molecules are complexed to quencher. The fluorescence intensity of solutions without M^+ are therefore lower, and smaller concentrations of M^+ can cause significant fluorescence increases (provided K_M is large enough to compete effectively with K_Q).

The calculations presented in the form of Figure 3 suggest that for the measurement of concentrations of M^+ of 10^{-7} molar or higher, crowns with a K_Q value of about 10^4 and an appropriate K_M value (between 10^2 and 10^6) would be useful. Higher values of K_Q would be counter productive. A crown with K_Q and K_M values of about 10^4 would be predicted to allow measurement of M^+ at about 10^{-5} moles per liter.

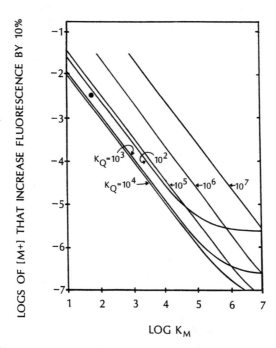

Figure 3. Calculated concentrations of M^+ (log $[M^+]$ plotted) that displace enough quencher to cause a 10% increase in fluorescence versus the log of the complexation constant for M^+ (K_M) at selected values of the complexation constant for the quencher, K_Q. The assumptions were: $[Q]_o = 1 \times 10^{-3}$ M, M^+ causes only 15% quenching when fully complexed, and Q causes 99.5% quenching when fully complexed. The dot corresponds to the observed K_Q and K_M values for crown 1.

Cation-Fostered Interruption of Quenching in a Crown Containing Both a Fluorescent Chromophore and a Quencher

Scheme III illustrates the concept in which complexation of a cation by the crown stiffens the crown ring and enforces a separation of the quencher and chromophore. Complexation may also help to tie up electron pairs that may act as electron-donor-acceptor quenchers. Two early examples in the literature were reported by Konopelski et al. (6) and de Silva et al. (11), and many cases of enhanced fluorescence are likely to have this concept as a basis. We synthesized crown ether 2, a simple molecule that we hoped would illustrate some of the principles of Scheme III (22). Crown 2 has a naphthalene fluorescent chromophore and an aniline electron donor quencher.

The synthesis of 2 was relatively straight-forward, and the yield in the crown forming synthetic step was 30 percent. Spectral and elemental analysis data support the structure assignment, and repeated recrystallization from hexane gave compound suitable for fluorescence studies.

Figure 4 shows fluorescence spectra of 2 in methanol in the absence (a) and presence (b) of potassium acetate. The peak at 340 nm is fluorescence from the naphthalene chromophore, and the broad peak in the 500 nm region is fluorescence from an excited electron donor acceptor (EDA) complex. As expected for fluorescence from an EDA complex, the energy and the intensity of the 500 nm band is increased as solvents are varied through the sequence of methanol, acetonitrile, dichloromethane, and cyclohexane. The energy changes are presumably due to stabilization of the polar excited EDA complex by polar solvents.

As envisioned in Scheme III the addition of potassium ions increases the intensity of the naphthalene fluorescence (and decreases the fluorescence from the EDA complex). The potassium ion also restores the vibrational structure of the fluorescence from the naphthalene chromophore and makes it similar to that of the naphthalene

CATION-FOSTERED INTERRUPTION OF QUENCHING IN A CROWN
CONTAINING BOTH A FLUORESCENT CHROMOPHORE AND A QUENCHER

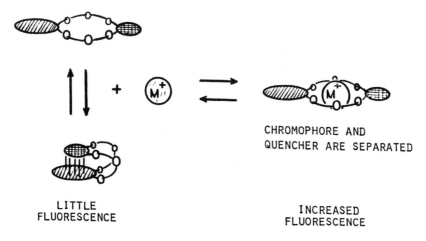

CHROMOPHORE AND
QUENCHER ARE SEPARATED

LITTLE
FLUORESCENCE

INCREASED
FLUORESCENCE

Scheme III.

chromophore in model crown ether **3**. The integrated naphthalene fluorescence intensity is increased by 55 percent when 0.16 molar K^+ is added as potassium acetate (results are nearly anion independent).

Three major factors govern the potential extent of cation-enhanced fluorescence of a given compound operating by Scheme III. They are: 1) the extent of intramolecular quenching (if there is not significant quenching, interrupting it will not result in significant enhancement of fluorescence); 2) the effectiveness of cation complexation at interrupting the quenching process; and 3) the magnitude of the complexation constant of the sensing reagent.

To determine the extent of intramolecular quenching in **2** we compared the integrated fluorescence intensity of **2** with that of model compound **3** which has the same size crown ring, but does not contain the aniline quencher. Concentrations were adjusted so the optical densities were identical (and low) at a wavelength at which the aniline chromophore of **2** absorbed at a minimum. The comparison of the fluorescence intensities indicates that the aniline quenches 90 percent of the naphthalene fluorescence. If all quenching were interrupted, the enhancement would be 1000 percent. To investigate the effectiveness of complexation at interrupting quenching, a fluorescence "titration" shown in Figure 5 was carried out, and the data was treated using an NLLSQ program. The value of I_{fK}, the fluorescence intensity that would be observed if every crown ether had one complexed K^+ ion, was calculated to be only 105 percent larger than that of the uncomplexed crown. Thus, complete complexation would only interrupt about 10 percent of the quenching in compound **2**. Finally, the

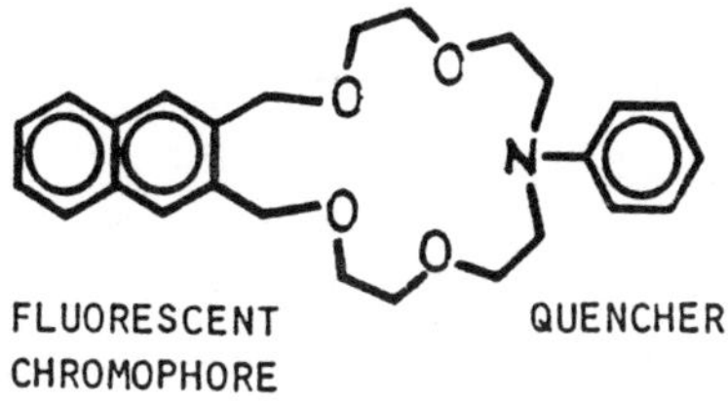

FLUORESCENT   QUENCHER
CHROMOPHORE
2

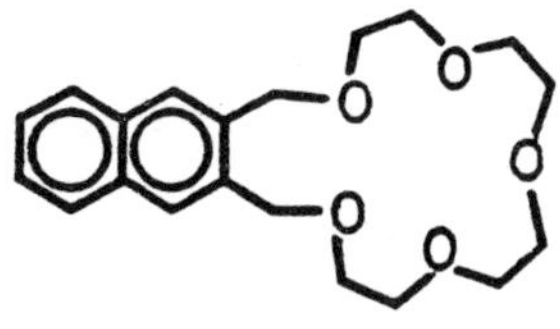

3

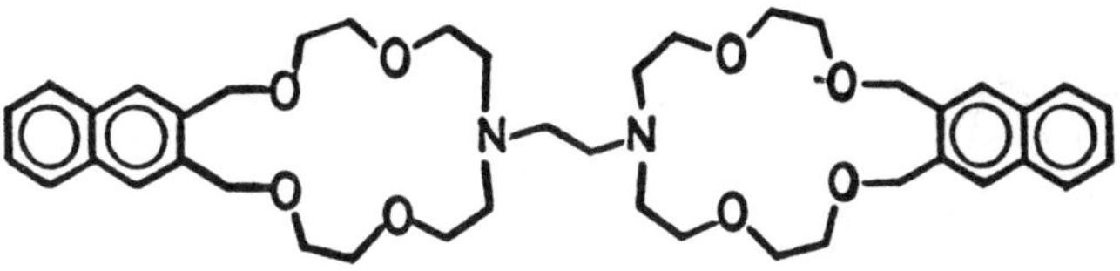

4

NLLSQ data treatment calculated a complexation constant of about 9. The low sensitivity of **2** toward K^+ is explained by very incomplete interruption of quenching by complexed K^+ and by the low complexation constant. Crown ether compounds that are likely to allow more complete interruption of quenching and have larger complexation constants are under construction.

Interruption of Intramolecular Quenching in a Bis(crown ether) Fluorescent Extraction Experiment

Bis(crown ether) compound **4** was synthesized in the hope that both crown ether rings would sandwich one metal ion between them, and that this process would bring the two naphthalene chromophores into close proximity so that an excited dimer (excimer) would form and a characteristic fluorescence band would appear. Early examples of metal ion fostered excimer formation were reported by Bouas-Laurent and co-workers(*7*) and Tundo and Fendler(*8*). Compound **4** has not shown excimer fluorescence, probably because the decrease in energy is small when two naphthalene chromophores form an excimer and because the 2,3-disubstitution on the naphthalene rings may present steric barriers to excimer formation. However, **4** does exhibit cation enhanced fluorescence as an extraction agent(*21*) in dichloromethane(*22*).

Figure 6 depicts the process of cation enhancement of fluorescence. Bis(crown ether) **4** at 2×10^{-5} molar concentration in the dichloromethane phase experiences intramolecular quenching. Using model compound **3** we determined that 75 percent of the fluorescence is quenched, presumably by intramolecular EDA quenching. When the metal ion in the water phase can be extracted by bis(crown ether) **4** into the dichloromethane phase, the complexation decreases the intramolecular quenching dramatically.

Two important factors must be taken into account as the extraction experiments are carried out. The extent of extraction is strongly dependent on the anion present. The hydroxide anion does not enable K^+ to be extracted into the dichloromethane phase so KOH in the water phase has no effect on the fluorescence of the dichloromethane phase. Chloride does enable K^+ to be extracted, but not extremely readily, so the

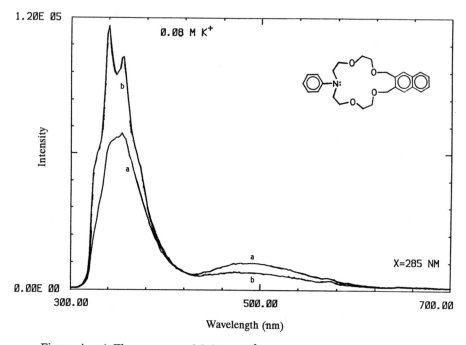

FLUORESCENCE SPECTRA IN METHANOL WITH ADDED KOAC

Figure 4. a) Fluorescence of 2.00 x 10^5 M crown **2** in methanol at room temperature (excitation at 285 nm). b) Fluorescence of **2** with added 0.08 M potassium acetate.

sensitivity is reduced compared to the case with the hexafluorophosphate anion. The large size and attendant low hydration make PF_6^- an effective anion for extraction experiments. To supply the needed mobile anion, experiments were done with 2.0 x 10^{-3} molar tetramethylammonium hexafluorophosphate in the water phase. The tetramethylammonium cation has a negligible effect on the fluorescence of **4**. Another complication is the competition among different cations that might complex with **4**. When PF_6^- is present in the water phase, hydronium ion reacts readily with **4**, protonating the nitrogen quencher and enhancing fluorescence by about 300 percent. A pH titration shows that the pK_a of **4** is approximately 10. To insure that observed enhancements of fluorescence are not due to protonation, all extraction experiments were run with the pH of the water phase adjusted to 11.5 using tetraethylammonium hydroxide.

Figure 7 shows the enhancement of fluorescence in the dichloromethane phase caused by KCl in the water phase (the two phases were shaken together for five minutes, and separated). As discussed above the aqueous pH was adjusted to 11.5, and the mobile anion PF_6^- was added to the aqueous phase. Figure 8 shows that the extraction process with **4** is selective for potassium ions over sodium ions by a factor of about 30, and for potassium ions over lithium or calcium ions by factors of about 300.

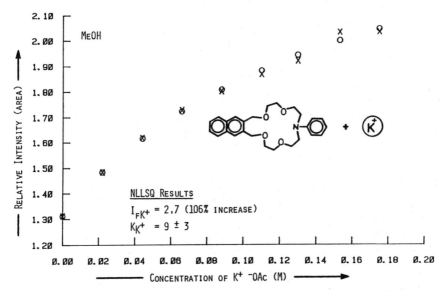

Figure 5. Relative fluorescence intensity (integrated over the naphthalene region) of crown 2 in room temperature methanol versus the concentration of added potassium acetate. Experimental values (x) and NLLSQ calculated values (o) are shown. As defined in the text, $[Cr]_o = 2.00 \times 10^{-5}$; and as determined by NLLSQ, $I_{fK} = 2.7$ (106% increase), and $K_K = 9$.

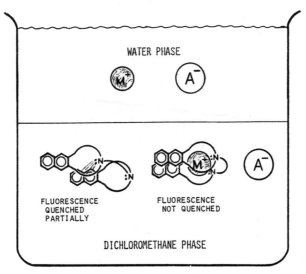

Figure 6. Cartoon illustrating the extraction experiment using the bis(crown ether) fluorescence sensing agent 4.

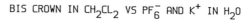

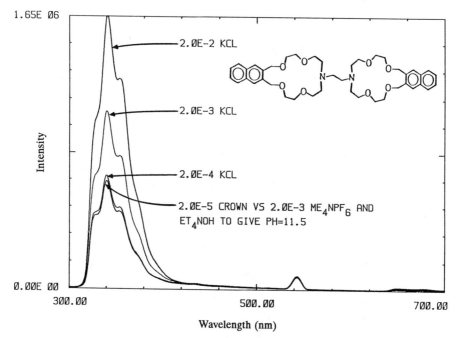

Figure 7. Fluorescence spectra of the dichloromethane phases containing 2.00 x 10^{-5} M crown **4** after equilibration with aqueous solutions with the concentrations of KCl shown on the Figure. All aqueous phases initially have 2.00 x 10^{-3} M tetramethylammonium hexafluorophosphate and pH's set to 11.5 using tetraethylammonium hydroxide.

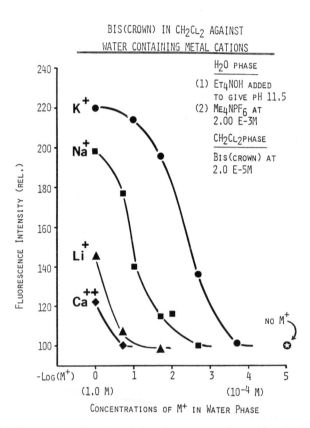

Figure 8. Curves showing the relative fluorescence intensities (at 340 nm) of dichloromethane phases containing bis(crown) **4** (at 2.00 x 10^{-5} M) plotted as a function of the concentration of metal ions (K^+, Na^+, Li^+, and Ca^{+2}) added to the aqueous phase (with added PF_6^- and pH = 11.5).

Acknowledgments

S.J.G., B.E.B, S.B. and S.P gratefully acknowledge the Ball State University Honors College Undergraduate Fellows Program, and T.E.M. gratefully acknowledges the Ball State University Undergraduate Research Grant Program. L.R.S. gratefully acknowledges the Ball State Chemistry Department Summer Research Program, the Ball State Faculty Grants Program, and the National Science Foundation (Grant No. CHE-8400 031).

Literature Cited

(1) Sousa, L. R.; Larson, J. M. *J. Am. Chem. Soc.* **1977**, *99*, 307.

(2) Larson, J. M.; Sousa, L. R. *J. Am. Chem. Soc.* **1978**, *100*, 1943.

(3) Shizuka, H.; Takada, K.; Morita, T. *J. Phys. Chem.* **1980**, *84*, 994.

(4) Wolfbeis, O. S.; Offenbacher, H. *Monatsh. Chem.* **1984**, *115*, 647.

(5) Grynkiewicz, G.; Poenie, M.; Tsien, R. Y. *J. Biol. Chem.* **1985**, *260*, 3440.

(6) Konopelski, J. P.; Kotzyba-Hibert, F.; Lehn, J.-M.; Desvergne, J.-P.; Fages, F.; Castellan, A.; Bouas-Laurent, H. *J. Chem. Soc., Chem. Commun.* **1985**, 433.

(7) Bouas-Laurent, H.; Castellan, A.; Daney, M.; Desvergne, J.-P.; Guinand, G.; Marsau, P.; Riffaud, M.-H. *J. Am. Chem. Soc.* **1986**, *108*, 315.

(8) Tundo. P.; Fendler, J. H. *J. Am. Chem. Soc.* **1980**, *102*, 1760.

(9) Haugland, R. P. *Molecular Probes: Handbook of Fluorescent Probes and Research Chemicals*; Molecular Probes, Inc.: Eugene, OR, 1992, pp 142-152.

(10) Moore, E. D. W.; Tsien, R. Y.; Minta, A.; Fay. F. S. *Fed. Am. Soc. Exp. Biol. J.* **1988**, *2(4)*, Abs. 2660.

(11) de Silva, A. P.; de Silva, S. A. *J. Chem. Soc., Chem. Commun.* **1986**, 1709.

(12) de Silva, A. P.; Sandanayake, K. R. A. S. *Tetrahedron Lett.* **1991**, *32*, 421.

(13) Fery-Forgues, S.; Le Bris, M.-T.; Guette, J.-P.; Valeur, B. *J. Chem. Soc., Chem. Commun.* **1988**, *5*, 384.

(14) Huston, M.; Haider, K.; Czarnik, A. W. *J. Am. Chem. Soc.* **1988**, *110*, 4460.

(15) Loehr, H.-G.; Voegtle, F. *Acc. Chem. Res.* **1985**, *18*, 65.

(16) Sanz-Medel, A.; Gomis, D. B.; Alvarez, J. R. G. *Talanta* **1981**, *28*, 425.

(17) Street, K. W. Jr.; Krause, S. A. *Anal. Lett.* **1986**, *19*, 835.

(18) Tran, C. D.; Zhang, W. *Anal. Chem.* **1990**, *62*, 835.

(19) Negulescu, P. A.; Harootunian, A.; Tsien, R. Y.; Machen, T. E. *Cell Regul.* **1990**, *1*, 259.

(20) Kina, K.; Shiraishi, K.; Ishibashi, N. *Bunseki Kagaku* **1978**, *27*, 291.

(21) Nishida, H.; Katayama, Y.;K, H.; Katsuki, H.; Nakamura, H.; Takagi, M.; Ueno, K. *Chem. Lett.* **1982**, 1853.

(22) Reported in part at the 190th American Chemical Society National Meeting, Chicago, Il, Sept. 13, 1985.

(23) Reported in part at the 197th American Chemical Society National meeting, Dallas, Tx, April 11, 1989.

(24) Christian, S. D.; Tucker, E. E. *Am. Lab.* **1982**, *14*

RECEIVED March 11, 1993

Chapter 3

Ion Recognition Detected by Changes in Photoinduced Charge or Energy Transfer

Bernard Valeur, Jean Bourson, and Jacques Pouget

Laboratoire de Chimie Générale, Conservatoire National des Arts et Metiérs, 292 rue Saint-Martin, 75003 Paris, France

Fluoroionophores consisting of a merocyanine (DCM), a styryl derivative of benzoxazinone and coumarin 153 linked to a macrocycle (aza-crown ether) undergo drastic changes in absorption and emissive properties under cation complexation. In these compounds, the crown acts as a tuner of the intramolecular charge transfer induced by light: the complexed cation enhances or decreases this charge transfer depending on its charge and its size with respect to the cavity size of the crown. The concomitant changes in the fluorescence properties can thus be used for cation recognition. Another original fluorescent sensor composed of two different coumarins linked by a flexible spacer (penta(ethylene oxide)) can bind efficiently lead(II); the resulting changes in the photophysical properties and, in particular, the efficiency of excitation energy transfer between the two coumarins, can be used for monitoring the complexation.

Molecular recognition is a subject of considerable interest because of its implications in many fields: biology, medicine (clinical biochemistry), environment, etc. In particular, the detection of metal cations involved in biological processes (e.g. sodium, potassium, calcium, magnesium), in clinical diagnostics (e.g. lithium, potassium, aluminium) or in pollution (e.g. lead, mercury, cadmium) has received considerable attention. Among the numerous methods employed, fluorescent sensors offer distinct advantages in terms of sensitivity and specificity. Moreover, remote sensing is possible by using optical fibres.

Our aim is to design fluorescent sensors that undergo photophysical changes (extinction coefficient, fluorescence intensity, shifts in absorption and emission spectra) as marked as possible upon cation binding. Special attention is of course paid to selectivity and complex stability. With this in mind, two types of systems are being investigated in our laboratory: *fluoroionophores* and *complexing bifluorophores*.

Fluoroionophores consist of a fluorophore (fluorescent molecule) covalently linked to an ionophore (e.g. crown ether) (Scheme 1). Complexation is expected to alter the photophysical properties of the fluorophore and these changes are used for the detection of ions.

Complexing bifluorophores are composed of two different fluorescent dyes linked by a flexible spacer containing heteroatoms (oxygen, nitrogen or sulphur atoms). Cation binding is thus possible and results in a decrease of the distance between the two fluorophores (Scheme 2). Therefore, as the emission spectrum of the donor (D) overlaps the absorption spectrum of the acceptor (A), an increase in the photoinduced energy transfer between the two moieties is to be observed.

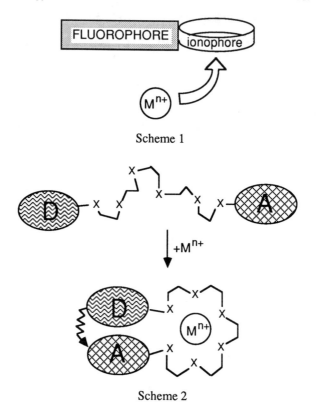

Scheme 1

Scheme 2

Fluoroionophores

Various crown-ether-linked fluorophores have been described (*1-14*). The choice of the fluorophore and the way the crown is linked to the latter are of major importance. We have selected the following fluorophores: a styryl derivative of benzoxazinone (BOZ-NMe2), a merocyanine (DCM, a well-known laser dye), and coumarin 153 (C153).

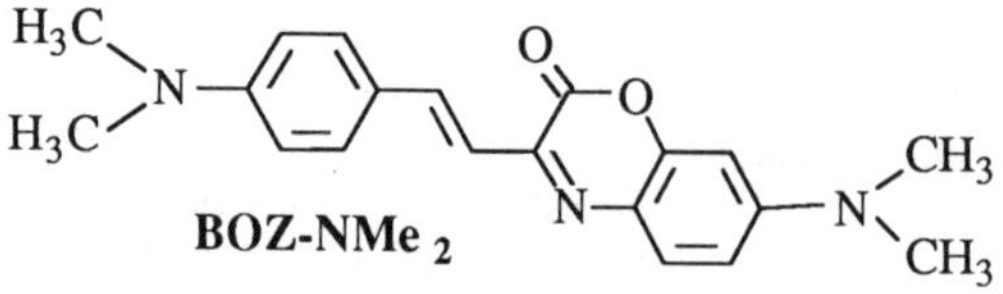

BOZ-NMe$_2$

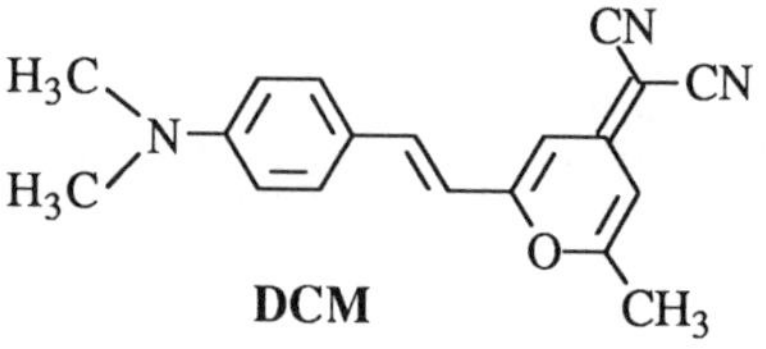

DCM

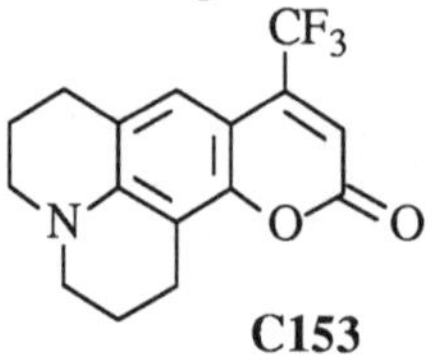

C153

The reason for these choices is the fact that upon light excitation they all undergo intramolecular charge transfer from an electron-donating group (amino group) to an electron-withdrawing group (dicyanomethylene group in DCM, heterocyclic nitrogen atom and carbonyl group in BOZ-NMe2, carbonyl group in coumarin 153). These fluorophores were chemically modified so that an aza-crown ether is incorporated in the structure to form the fluoroionophores BOZ-crown (*8,9,11*), DCM-crown (*10,11,20*), C153-crown (*14,15*) and (C153)2-K22 (*15*) (Figure 1). The crown plays the role of a *tuner* of the photoinduced charge transfer because complexation by cations enhances or reduces the photoinduced charge transfer. As a matter of fact, cation binding reduces the electron-donating character of the nitrogen atom of the crown in DCM-crown and BOZ-crown, whereas the complexed cation enhances the electron-withdrawing character of the carbonyl group of the crowned coumarins. The resulting changes in the photophysical properties are expected to depend on the cation size (as compared to that of the crown cavity) and on its charge.

The main features of these fluoroionophores are described herafter. They show the extraordinary variety of the photophysical effects that can be induced by cation binding.

BOZ-crown (*8,9,11*). The changes in the absorption and emission spectra upon complexation are shown in Figure 2. The absorption spectrum is blue shifted and the extinction coefficient is reduced. This can be interpreted as follows. The complexed cation attracts more or less (according to its charge density) the lone pair of the nitrogen atom of the crown, and therefore it reduces the conjugation in the molecule; an antiauxochromic effect (blue shift and hypochromic effect) is thus observed.

The same explanation holds for the blue shifts observed for the fluorescence spectra. As for the absorption spectra, the shifts are larger for divalent cations which are more efficient than monovalent ones in reducing the electron-donating character of the nitrogen atom belonging to the crown. On the other hand, the interpretation of the enhancement of the fluorescence intensity requires further attention. Comparison of BOZ-crown (either free or complexed by cations) with BOZ-NMe2 and the uncrowned compound BOZ-H is very helpful.

The fluorescence quantum yields (Φ_F) and lifetimes (τ_F) were determined and allowed us to calculate the rate constants for radiative and nonradiative processes according to the following relations:

$$k_r = \Phi_F / \tau_F \qquad\qquad k_{nr} = (1-\Phi_F) / \tau_F$$

The results are given in Table I. It is remarkable that the radiative rate constants k_r of the free ligand, all the complexes and the model compounds are found to be very similar. In contrast, marked changes in the value of the nonradiative rate constant k_{nr} are to be noted. k_{nr} is the sum of all possible modes of deexcitation: internal conversion, intersystem crossing and twisting out of the planar geometry in the excited state. Intersystem crossing due to spin orbit coupling is small because we have observed that a heavy atom like Ba^{2+} has weak quenching effects on BOZ-Me2; in addition, after total complexation of BOZ-crown by Ba^{2+}, a plateau for fluorescence intensity is observed over a wide range of barium concentrations.

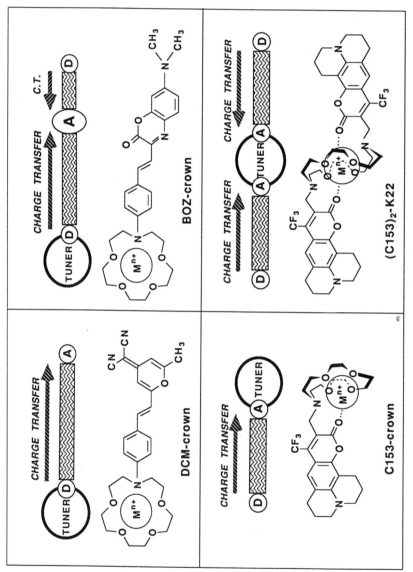

Figure 1. Cation control of photoinduced charge transfer in fluoroionophores.

Table I. Emissive properties of BOZ-crown and its complexes in acetonitrile. Stability constants K_S (Adapted from ref. 9)

	λ_F (nm)	Φ_F	τ_F (ns)	k_r (10^8 s^{-1})	k_{nr}	$\log K_S$	ionic diameter (Å)	
BOZ-crown[a]	642	0.33	2.1	1.6	3.2			
Li$^+$ $\subset$ BOZ-crown	612	0.44	2.8	1.6	2.0	2.83	1.36	Li$^+$
Na$^+$ $\subset$ BOZ-crown	611	0.38	2.5	1.5	2.5	2.23	1.94	Na$^+$
Mg^{2+} $\subset$ BOZ-crown	573	0.48	3.3	1.5	1.6	2.68	1.32	Mg^{2+}
Ca^{2+} $\subset$ BOZ-crown	574	0.64	3.3$_5$	1.9	1.1	4.14	1.98	Ca^{2+}
Ba^{2+} $\subset$ BOZ-crown	576	0.59	3.0$_5$	1.9	1.3	3.62	2.68	Ba^{2+}
BOZ-H	567	0.54	4.0	1.4	1.1			
BOZ-NMe$_2$	649	0.23	1.6	1.4	4.8			

[a]Cavity size of the crown : 1.7 - 2.2 Å.

Therefore, twisting out of the planar geometry in the excited state should explain the differences in the values of k_{nr}. The following order is observed:

$$\text{BOZ-H} \ll \text{BOZ-crown} < \text{BOZ-NMe}_2$$

The complexes have intermediate k_{nr} values between those of BOZ-H and BOZ-crown. The k_{nr} value of the calcium complex is very close to that of BOZ-H because Ca^{2+} almost monopolizes the lone pair of the nitrogen atom and therefore the behavior of this complex approaches that of the uncrowned compound BOZ-H. The differences observed between complexes with various cations can thus be interpreted in terms of ability of the cations to hinder the electronic effects of the nitrogen atom of the crown, and thus to mask the presence of the crown.

The exact nature of twisting out of the planar geometry is difficult to determine. A simple process of energy dissipation due to rotation of the crown can be invoked (9), but the existence of a TICT (twisted internal charge transfer) state formed by single bond twisting upon excitation is likely in this kind of molecule. In fact, CNDO/S calculations show that BOZ-NMe$_2$ possesses a low-lying TICT state (twisted single bond connecting benzoxazinone and ethylene) (16).

The size of the cations with respect to the cavity size of the monoaza-15-crown-5 moiety is now to be considered. Amongst cations of the same charge, the largest effects are observed for those cations whose ionic diameter (see Table I) is closest to the cavity size i.e. 1.7 to 2.2 Å. Moreover a good correlation with the stability constants of the complexes is noted (see Table I). A noticeable exception concerns the lithium ion: the stability constant of the lithium complex is higher than that of the sodium complex despite that sodium ion should fill out more optimally the ligand cavity. A possible explanation of this discrepancy is that the lithium ion is amongst those alkaline ions the most strongly solvated in acetonitrile. Therefore a better fit to the slightly larger cavity should be obtained with the solvated lithium ion, as observed in water (17).

DCM-crown (10,11,20)). It is generally admitted that the fluorescence emission of DCM and its derivatives results almost solely from an intramolecular charge transfer

(ICT) from the donor moiety (amino group) to the acceptor moiety (dicyanomethylene group) (*18,19*). The polarity of the solvent plays an important role in the stabilization of this ICT state, as revealed by the strong solvatochromism resulting from the large increase of the dipole moment upon excitation. The photophysical properties of DCM-crown which are very similar to those of DCM are affected by cation binding (*10*). As shown in Figure 3, cation binding induces an antiauxochromic effect upon the absorption spectrum but much more pronounced than for BOZ-crown. Again this effect is much more marked for cations of higher charge density. It is interesting to note the similarity between the absorption spectrum of the reference compound DCM-H (which does not have the auxochromic amino group) and those of the fully complexed ligand with Ca^{2+} and Mg^{2+} (Figure 3).

DCM-H

On the other hand, the slight changes in position and shape of the fluorescence spectra upon cation binding are surprising (Figure 3). Moreover, the fluorescence lifetime is almost unaffected whereas the fluorescence quantum yield is reduced (Table II). These observations are in contrast with the drastic changes of the absorption spectra. A plausible explanation could be the photoejection of the cation. Photoejection can be explained as follows: owing to the photoinduced charge transfer from the nitrogen atom of the crown to the dicyanomethylene group, the nitrogen of the crown becomes positively polarized, and the resulting repulsion with the complexed cation causes photoejection of the cation. Direct evidence of photoejection of lithium and calcium ions is provided by picosecond pump-probe spectroscopy (*20*) which shows that the photoejection process takes place in less than 5 and 20 ps for the

Table II. Emissive properties of DCM-crown and its complexes in acetonitrile. Stability constants K_S

	λ_{abs} (nm)	λ_F (nm)	Φ_F	τ_F (ns)	$log\ K_S$	ionic diameter (Å)	
DCM-crown[a]	464	621	0.73	2.20			
$Li^+ \subset$ DCM-crown	440	606	0.57	1.88	2.48	1.36	Li^+
$Na^+ \subset$ DCM-crown	427	611	0.65	2.06	1.98	1.94	Na^+
$Mg^{2+} \subset$ DCM-crown	394	610	-	2.10	-	1.32	Mg^{2+}
$Ca^{2+} \subset$ DCM-crown	398	608	0.27	1.94	3.75	1.98	Ca^{2+}
$Ba^{2+} \subset$ DCM-crown	401	611	0.40	1.99	2.98	2.68	Ba^{2+}
DCM	460	627	0.60	1.90			
DCM-H	391	490	3×10^{-4}	-			

[a]Cavity size of the crown : 1.7 - 2.2 Å.

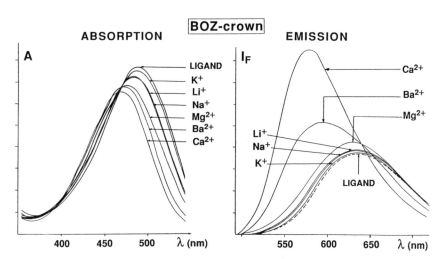

Figure 2. Absorption and emission spectra of BOZ-crown in acetonitrile before and after addition of alkali or alkaline-earth metal perchlorate. (Absorption spectra: the concentration is 2×10^{-5} M in ligand and 2×10^{-3} M in perchlorate salts; Emission spectra: the concentration is 1.5×10^{-6} M in ligand and 2×10^{-4} M in perchlorate salts; spectra are corrected according to the absorbance of the solutions at the excitation wavelength of 480 nm). (Adapted from ref. 9).

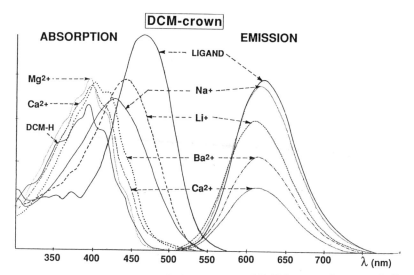

Figure 3. Absorption and emission spectra of DCM-crown in acetonitrile before and after full complexation by alkali or alkaline-earth metal ions. (Adapted from ref. 10).

Li^+ and Ca^{2+} complexes, respectively. This type of photoejection may provide in the future a new way for the photogeneration of cations which is of interest for the study of the response of some chemical or biological systems to a jump of cation concentration.

The stability constants of the complexes with DCM-crown in acetonitrile are given in Table II. They follow the same order as those of BOZ-crown but they are lower by a factor of 2-2.5. In order to interpret this systematic difference, we studied by NMR and optical spectroscopy the binding capacity of N-phenylaza-15-crown-5 ether linked to various substituents X (NO, CHO, H, CH_2OH, NH_2) in the para position to the crown (11). The stability constants of the complexes with various alkali-metal and alkaline-earth-metal ions were determined by spectrophotometry (see Table III for the calcium complexes). Upon complexation, the variations in the NMR chemical shift δ_A of the protons in the ortho position to the crown parallel the stability constants of the complexes. The chemical shift δ_X of the protons in the ortho position to the substituent X reflects the electron-withdrawing or -donating character of this substituent, but δ_X is much less sensitive to the perturbation of the nitrogen atom of the crown upon complexation. Consequently, the internal chemical shift $\delta_X - \delta_A$ which in principle reflects the basicity of this nitrogen atom, and thus its complexing ability, is not well correlated with the stability constant when the substituent is varied. On the other hand, there is a satisfactory correlation between the stability constant and the variation Δ in internal chemical shift defined as :

$$\Delta = (\delta_X^L - \delta_A^L) - (\delta_X^{ML} - \delta_A^{ML})$$

where the superscripts L and ML refer to the ligand and the complex, respectively. The high stability constants of the complexes with fluoroionophores (i.e., in the case where the substituent X extends the conjugation of N-phenylaza-15-crown-5) cannot be simply interpreted in terms of the more or less electron-withdrawing character of this substituent, but other effects are involved.

Table III : Spectral characteristics, stability constants K_S and variation Δ in internal chemical shift of calcium complexes of 4-substituted phenylaza-15-crown-5 ethers in acetonitrile (Adapted from ref. 11)

X	λ_{abs} (nm)	ε ($M^{-1}cm^{-1}$)	$log\,K_S$	Δ^a (ppm)
NO	419	30600	1.99	0.07
CHO	336	34300	2.59	0.12
H	303	2300	3.23	0.06
CH_2OH	306	1300	3.06	0.21
NH_2	331	1400	3.44	0.34
DCM	464	49300	3.72	0.38
BOZ	490	38200	4.14	0.41

$^a\Delta = (\delta_X^L - \delta_A^L) - (\delta_X^{ML} - \delta_A^{ML})$

C153-crown (*14,15*) **and C(153)₂-K22** (*15*). In C153-crown, the crown is the same as in BOZ-crown and DCM-crown, but in contrast to the latter two the nitrogen atom of the crown is not involved in the conjugation of the dye. Therefore, the changes in photophysical properties upon complexation are due to direct interaction between the cation and the carbonyl group of the coumarin (Figure 1). Upon excitation, there is a charge transfer from the amino group to the carbonyl group, and the complexed cation enhances this charge transfer. Red shifts of the absorption and emission spectra are thus observed (Figure 4). This shift follows exactly the order of increased charge density of the cation.

The changes in quantum yield of C153-crown (Table IV) can be interpreted as follows. In the free ligand, the crown can interact with the dye which results in some quenching. Upon complexation these interactions disappear which leads to an increase in quantum yield; this is indeed observed with Na^+, Ba^{2+} and Li^+. But the decrease in quantum yield upon complexation with Ca^{2+} and Mg^{2+} shows that there is an additional competitive route of deexcitation which seems to be related to the charge density of the cation. Evidence for interaction between the crown and the free ligand is provided by the decay of fluorescence which is not monoexponential, whereas all the complexes exhibit a monoexponential decay.

Table IV. Fluorescence quantum yield of C153-crown, (C153)₂-K22 and their complexes (Adapted from ref. 15)

	ligand	Complexes					
		Li^+	Na^+	K^+	Mg^{2+}	Ca^{2+}	Ba^{2+}
Ionic diameter (Å)		1.36	1.94	2.66	1.32	1.98	2.68
C153-crown[a] Φ_F	0.15	0.16	0.19	0.19	0.11	0.12	0.16
(C153)₂-K22[b] Φ_F	0.016		0.017	0.04	0.011	0.009	0.10

[a]Cavity size of the crown : 1.7 - 2.2 Å.
[b]Cavity size of the crown : 2.6 - 3.2 Å.

The absorption and emission spectra of (C153)₂-K22 are displayed on Figure 5. The quantum yield of the free ligand is ten times lower than that of C153-crown; self-quenching is indeed possible because the flexibility of the crown allows the two coumarin moieties to come in contact. Upon complexation there is an increase in quantum yield only with Ba^{2+} and K^+, and these cations are precisely those which fit very well the cavity of the crown, whereas the others are smaller (Table IV). Therefore, it can be anticipated that in the complexes with Ba^{2+} and K^+ the carbonyl groups of the two coumarins are preferentially on opposite sides with respect to the cation (as shown in Figure 1) and that self-quenching is thus totally or partially suppressed. On the other hand, since the other cations are smaller than the cavity size of the crown (Table IV), the preferred conformation of the relevant complexes may

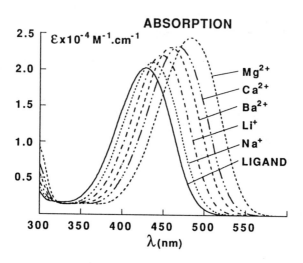

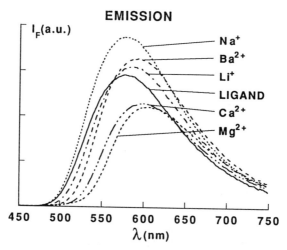

Figure 4. Absorption and emission spectra C153-crown in acetonitrile before and after full complexation by alkali or alkaline-earth metal perchlorate. (Adapted from ref. 15).

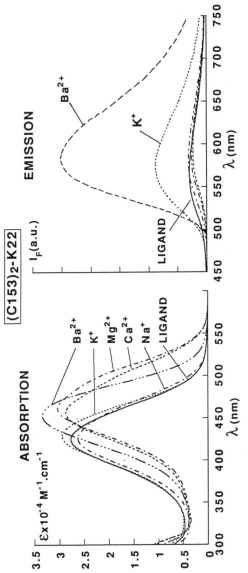

Figure 5. Absorption and emission spectra of(C153)2-K22 in acetonitrile before and after full complexation by alkali or alkaline-earth metal perchlorate. (Adapted from ref. 15).

be such that the two carbonyls are on the same side; the resulting close distance between the two coumarin moieties (and possibly stacking) accounts for quenching. The existence of such a folded conformation is supported by the absorption spectra of the complexes with Na^+, Ca^{2+} and Mg^{2+} which exhibit shoulders on the red edge side arising from ground-state interactions (Figure 5). In contrast, no shoulder is observed in the absorption spectra of the complexes with Ba^{2+} and K^+ because of the absence of ground-state interactions in the extended conformations. X-ray diffraction measurements will be performed in order to get further insight in the conformation of the various complexes.

The stability constants of the complexes of C153-crown and $(C153)_2$-K22 are much higher than those of BOZ-crown and DCM-crown; this arises from the participation of the coumarin carbonyl group in the complexes and from the deconjugation of the nitrogen atom of the crown from the π-electron system. The stability constants in acetonitrile are given in Table V. They depend of course on the nature of the solvent. In methanol the stability constants are lower but still high for the coumarin fluoroionophores; for example, the stability of the complex of $(C153)_2$-K22 with Ba^{2+} is 2.5×10^5 M^{-1}.

Table V. Stability constants of the complexes in acetonitrile

cation	Li^+	Na^+	K^+	Mg^{2+}	Ca^{2+}	Ba^{2+}
ionic dia-meter (Å)	1.36	1.94	2.66	1.32	1.98	2.68
BOZ-crown[a]						
log K_S	2.8	2.2	2.2	2.7	4.1	3.6
DCM-crown[a]						
log K_S	2.5	2.0			3.7	3.1
C153-crown[a]						
log K_S	6.3	5.0	4.2	6.2	6.8	6.7
(C153)$_2$-K22[b]						
log K_S			4.6	6.2	7.8	7.6

[a]Cavity size of the crown : 1.7 - 2.2 Å.
[b]Cavity size of the crown : 2.6 - 3.2 Å.

The complex stability of the coumarin fluoroionophores remains high enough in acetonitrile-water mixtures for practical applications to the determination of cation in aqueous samples. In order to examine the ability of C153-crown to detect cations, the following method has been employed. A stock solution of C153-crown in acetonitrile at a concentration of 8×10^{-6} M (absorbance of 0.160 at 428 nm) is prepared. 30 μl of water is added to 3 ml of this solution. The fluorescence intensity I_0 of this reference solution is measured upon excitation at 480 nm and observation at 584 nm. Under exactly the same experimental conditions, the fluorescence intensity I is measured with solutions obtained by adding 30 μl of water with various amounts of cation perchlorate to 3 ml of the stock solution. Let c_w and c be the concentrations of cation in water before addition, and in the cuvette after addition, respectively. The ratio I/I_0 is then plotted as a function of cation concentration (Figure 6). These calibration curves can be used for the determination of an unknown concentration of one of the cations in an aqueous neutral solution in the range 0.1-1 mM. As regards selectivity, the charge of the cation is the main factor that governs the efficiency of complexation; the larger the charge of the cation, the higher is the stability constant of the complex. The size of the cation with respect to the cavity size of the monoaza-15-crown-5 moiety also plays a role; however among cations of the same charge, the selectivity is not very good because of the flexibility of the link between the crown and the coumarin moiety bearing the carbonyl group.

In contrast to the fluorescent complexing agents of the EDTA type, which are well suited to the determination of cations (e.g. calcium) in the micromolar range pertaining to living cells (*21*), the fluoroionophores C153-crown and (C153)$_2$-K22 offer interesting characteristics for cation titration in the millimolar range, i.e., for biomedical applications (e.g., plasma analysis).

Complexing bifluorophores (*22,23*).

The bichromophoric molecule DXA consisting of two coumarins linked by a penta(ethylene oxide) spacer can bind Pb^{2+} ions in acetonitrile and in propylene carbonate.

DONOR   ACCEPTOR

DXA

The spectrocopic models of the coumarin donor and acceptor are 7-ethoxycoumarin and 4-trifluoromethyl-7-ethylaminocoumarin (coumarin 500), respectively. The emission spectrum of the coumarin-donor strongly overlaps the absorption spectrum of the coumarin-acceptor so that nonradiative resonance energy transfer can occur with a rate and an efficiency that depend of the distance between the two moieties. In previous papers, we reported the synthesis of this bifluorophore and studies of the distribution of interchromophoric distances by means of conformational calculations, and steady-state and time-resolved energy transfer experiments (*24,25*).

Lead perchlorate (Pb(ClO$_4$)$_2$, 3H$_2$O) was gradually added to a solution of DXA in acetonitrile and propylene carbonate containing a supporting electrolyte (tetraethylammonium perchlorate 0.1M). The changes in absorption spectrum are shown in Figure 7. The acceptor band (maximum at 385 nm) undergoes a blue shift and a hypochromic effect. This can be explained by the fact that, in the complex, Pb^{2+} attracts the lone pair of the nitrogen atom and thus reduces its electron-donating character (antiauxochromic effect). In contrast, there is no shift of the donor absorption band (maximum at 321 nm) but an increase in absorbance. This increase is only due to the shift of the acceptor band because a compound, consisting of the coumarin-donor linked to the spacer but without the acceptor moiety, exhibits no significant change in absorption upon addition of Pb^{2+}.

The emission and excitation spectra are given in Figure 8. The emission band of the donor moiety cannot be seen because the fluorescence quantum yield of 7-ethoxycoumarin (which is a good spectroscopic model for this moiety) is very low (8x10^{-3} in acetonitrile and 0.015 in propylene carbonate at 20°C) and even lower in

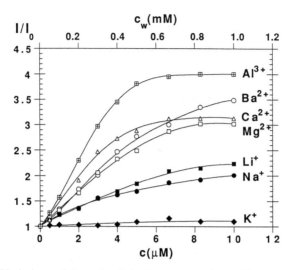

Figure 6. Variations of the ratio of the fluorescence intensities measured in the presence and in the absence of cation as a function of cation concentration. c_w and c are the concentrations of cation in water before addition and in the cuvette after addition, respectively. The concentration in C153-crown is 8×10^{-6} M. (Reproduced with permission from ref. 14. Copyright 1992 Elsevier Science Publishers).

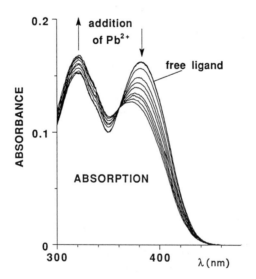

Figure 7. Changes in absorption spectrum of DXA induced by the addition of lead perchlorate. Solvent: acetonitrile containing tetraethylammonium perchlorate (0.1M) as a supporting electrolyte. The concentration in DXA is 1.0×10^{-5} M. Range of concentration in Pb^{2+}: 0 - 5×10^{-4} M. Temperature: 20 °C. (Adapted from ref. 23).

the presence of energy transfer. Upon excitation at 321 nm, i.e. in the donor absorption band, addition of Pb^{2+} induces a blue shift and an enhancement of the acceptor emission band. The explanation of the blue shift of the acceptor absorption band still holds for the emission band. As regards the enhancement, it is due to the increase in absorbance at 321 nm, and to an increase in transfer efficiency as a result of the conformational changes of the bichromophoric molecule induced by complexation with Pb^{2+}.

The transfer efficiency Φ_T, defined as the fraction of excited donors that transfer their energy to the linked acceptors, can be determined by comparison of the absorption and excitation spectra of the bichromophore at two wavelengths corresponding to the absorption maxima of the donor and the acceptor moieties, respectively (26). The values of Φ_T are reported in Table VI together with the Förster critical radii R_0 (distance at which transfer and spontaneous decay of the excited donor are equally probable, or distance at which $\Phi_T = 0.5$). There is only a small increase of Φ_T in acetonitrile and a very slight change in propylene carbonate. At first sight, such a slight change is surprising because complexation is expected to reduce the distance between the donor and acceptor moieties and thus to increase the efficiency of transfer.

Table VI. Transfer efficiencies Φ_T and Förster radii R_0 of the free ligand and complex in acetonitrile and in propylene carbonate (PC) (Adapted from ref. 24)

Solvent		Φ_T	$R_0 (Å)$
CH_3CN	free ligand	0.77	19.3
	complex	0.89	18.3
PC	free ligand	0.86	21.6
	complex	0.89	20.0

Excite-and-probe experiments (23) showed that the dynamics of transfer is not significantly affected by complexation in propylene carbonate, whereas faster transfer kinetics is observed for the complex in acetonitrile. This is consistent with the results on transfer efficiency.

Information on the structure of the complex is of great importance for the interpretation of the above observations. The stoichiometry of the complex can be determined from the changes in photophysical properties induced by complexation. It is convenient to monitor the changes in absorbance at 385 nm which are shown in Figure 9 and 10. In acetonitrile (Figure 9) the stoichiometry of the complex was found to be 1:1 and the stability constant 3.1×10^4 M^{-1}; it should be emphasized that this value is quite high and comparable to those observed with macrocyclic ligands. It is interesting to note that the stoichiometry of the complex is 1:1 but the transfer efficiency is not close to 1; therefore, the donor and acceptor moieties are not very close to each other, and helical wrapping of the ligand around one cation may explain the observations. Examples of such helical wrapping of acyclic ligands can be found in the literature (27). In contrast to acetonitrile, the sigmoidal shape of the titration curve observed in propylene carbonate (Figure 10) is remarkable and characteristic of cooperative binding. The best fit corresponds to a stoichiometry of 1:3 (ligand: metal ion) and the stability constant is 7.5×10^{11} M^{-3}. The structure of the complex is likely to be helical, as often observed with oligoethylene glycol ethers (27). Such a structure

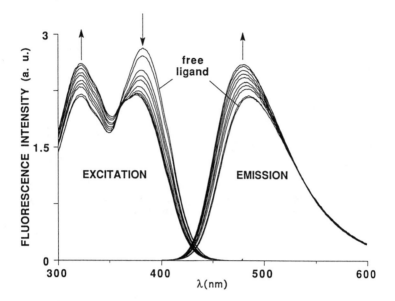

Figure 8. Emission and excitation spectra of DXA under the same conditions as those indicated in Figure 7. The excitation spectra are recorded at an observation wavelength of 480 nm. The concentration in DXA is 1.0×10^{-5} M. (Adapted from ref. 23).

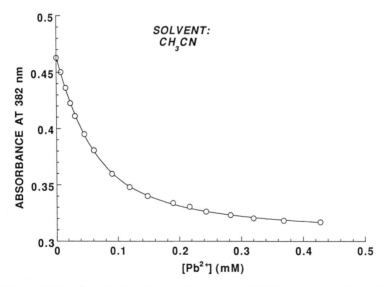

Figure 9. Variations in absorbance of a solution of DXA in acetonitrile at 382 nm as a function of lead concentration (o). The concentration of DXA is 2.5×10^{-5} M. The solid line represents the best fit corresponding to a stoichiometry of the complex of 1:1. The correlation coefficient is 0.99985. (Reproduced with permission from ref. 23. Copyright 1992 American Chemical Society).

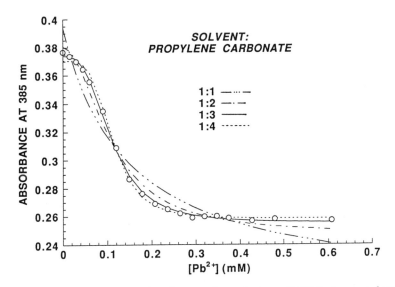

Figure 10. Variations in absorbance of a solution of DXA in propylene carbonate at 385 nm as a function of lead concentration (o). The concentration of DXA is 2.1×10^{-5} M. The solid line represents the best fit corresponding to a stoichiometry of the complex of 1:3. The correlation coefficients is 0.9993. Fits with other stoichiometries (1:1, 1:2, 1:4) are less satisfactory (Reproduced with permission from ref. 23. Copyright 1992 American Chemical Society).

is consistent with a slight change of transfer efficiency, because the spacer is very flexible (25) and the average center-to-center distance in the free ligand may not be very different from that in the complex which has an elongated form.

The difference in the structure and stoichiometry of the complexes in acetonitrile and propylene carbonate may be explained by the possibility of propylene carbonate molecules participating in the complexation through their carbonyl group. Water of crystallization of the salt and the perchlorate anion may also participate, whereas the carbonyl groups of the coumarins do not seem to be in a favorable position to be involved. Tentative schematic models of the complexes are shown in Figure 11. Comment should be made on the size of Pb^{2+}: its ionic radius is 1.18 Å, but it should be noted that Pb(II) probably has an active lone pair, and as such behaves as a very much smaller ion (28).

As regards the effects of other cations, it was found that the changes in photophysical properties upon addition of perchlorate salts of Li^+, Na^+, K^+, Cs^+, Ca^{2+}, Mg^{2+}, Ba^{2+} are very weak or insignificant owing to the very weak stability of the complexes.

Complexation of a bifluorophore would lead to a much larger increase in transfer efficiency if the interchromophoric distance were larger than the Förster critical radius in the free ligand and lower in the complex. In fact, because of the sixth power of the ratio R/R_0 involved in the expression of transfer efficiency (29)

$$\Phi_T = \frac{1}{1 + [\frac{R}{R_0}]^6}$$

Φ_T varies drastically around R/R_0 (Figure 12). Therefore, a large change in transfer efficiency could be in principle observed with a relatively small variation in distance.

Complexing bifluorophores are of limited use for the detection of metal cations for which many other methods are available, but they are of potential use for organic cations like guanidinium ion, as shown in scheme 3.

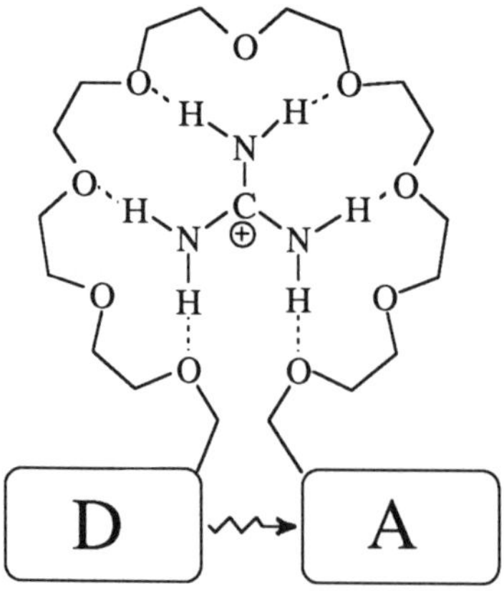

Scheme 3

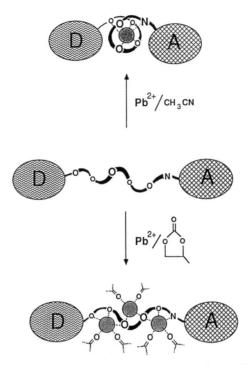

Figure 11. Tentative schematic illustration of the supramolecular complexes formed in acetonitrile and propylene carbonate. D is the coumarin donor and A the coumarin acceptor. The carbonyl groups of some propylene carbonate molecules are also shown. (Reproduced with permission from ref. 23. Copyright 1992 American Chemical Society).

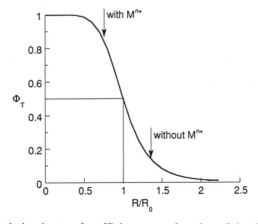

Figure 12. Variation in transfer efficiency as a function of the donor-acceptor distance. The arrows indicate favorable average distances before and after complexation for the tuning of transfer efficiency in a broad range. (Reproduced with permission from ref. 23. Copyright 1992 American Chemical Society).

Literature cited

1. Löhr, H. G.; Vögtle, F. *Acc. Chem. Res.* **1985**, *18*, 65.
2. Nishida, H.; Katayama, Y.; Katsuki, H.; Nakamura, H.; Takagi, M.; Ueno, K. *Chem. Lett.* **1982**, 1853.
3. Nakashima, K.; Nakatsuji, S.; Akiyama, S.; Tanigawa, I.; Kaneda, T.; Misumi, S.*Talanta*, **1984**, *31*, 749.
4. Nakashima, K.; Nagaoka, Y.; Nakatsuji, S.; Kaneda, T.; Tanigawa, I.; Hirose, K.; Misumi S.; Akiyama, S. *Bull. Chem. Soc. Jpn.*, **1987**, *60*, 3219.
5. Shinkai, S.; Miyazaki, K.; Nakashima M.; Manabe, M. *Bull. Chem. Soc. Jpn.*, **1985**, *58* , 1059.
6. Konopelski, J. P. ; Kotzyba-Hibert, F. ; Lehn, J. M. ; Desvergne, J. P. ; Fagès, F.; Castellan, A.; Bouas-Laurent, H. *J. Chem. Soc., Chem. Commun.* **1985**, 433.
7. De Silva , A. P. ; De Silva, S. A. *J. Chem. Soc., Chem. Commun.* **1986**, 1709.
8. Fery-Forgues, S.; Le Bris, M. T.; Guetté, J. P.; Valeur, B. *J. Chem. Soc. Chem.Commun.* **1988**, 384.
9. Fery-Forgues, S.; Le Bris, M. T.; Guetté, J. P.; Valeur, B. *J. Phys. Chem.*, **1988**, *92*, 6233.
10. Bourson, J.; Valeur, B. *J. Phys. Chem.* **1989**, *93*, 3871.
11. Fery-Forgues, S.; Bourson, J.; Dallery, L.; Valeur, B. *New J. Chem.* **1990**, *14*, 617.
12. Fages, F.; Desvergne, J. P.; Bouas-Laurent, H.; Lehn, J. M.; Konopelski, J. P.; Marsau, P.; Barrans, Y. *J. Am. Chem. Soc.* **1989**, *111*, 8672.
13. Desvergne, J. P.; Fages, F.; Bouas-Laurent, H.; Marsau, P. *Pure Appl. Chem.* **1992**, *64*, 1231.
14. Bourson, J.; Borrel, M. N.; Valeur, B. *Anal. Chim. Acta* **1992**, *257*, 189.
15. Bourson, J.; Pouget, J.; Valeur, B. *J. Phys. Chem.* **1993**, *97*, 4552.
16. Fery-Forgues, S.; Le Bris, M. T.; Mialocq, J. C.; Pouget, J.; Rettig, W.; Valeur, B. *J. Phys. Chem.*, **1992**, *96*, 701.
17. Pacey, G. E.; Wu, Y. P. *Talanta* **1984**, *31*, 165.
18. Meyer, M.; Mialocq, J. C. *Opt. Commun.* **1987**,*64*, 264.
19. Rettig, W.; Majenz, W. *Chem. Phys. Letters.* **1989**, *154*, 335.
20. Martin, M.; Plaza, P.; Dai Hung, N.; Meyer, Y.; Bourson, J.; Valeur, B. *Chem. Phys. Letters.* **1993**, *202*, 425.
21. Grynkiewicz, G.; Poenie, M.; Tsien, R. *J. Biol. Chem.* **1985**, *260*, 3440.
22. Valeur, B.; Bourson, J.; Pouget, J. *J. Lumin.* **1992**, *52*, 345.
23. Valeur, B.; Bourson, J.; Pouget, J.; Kaschke, M.; Ernsting, N. P. *J. Phys. Chem.* **1992**, *96*, 6545.
24. Valeur, B.; Mugnier, J.; Pouget, J.; Bourson, J.; Santi, F. *J. Phys. Chem.*, **1989**, *93*, 6073.
25. Kaschke, M.; Valeur, B.; Bourson, J.; Ernsting, N. P. *Chem. Phys. Lett.*, **1991**, *179*, 544.
26. Mugnier, J. , Pouget, J.; Bourson, J.; Valeur, B. *J. Lumin.*, **1985**, *33*, 273.
27. Vögtle, F.; Weber, E. *Angew. Chem. Int. Ed. Engl.*, **1979**, *18*, 753.
28. Hancock, R. D.; Shaikjee, M. S.; Dobson, S. M.; Boeyens, J. C. A. *Inorg. Chim. Acta.* **1988**, *154*, 229.
29. Förster, Th. *Z. Naturforsch.*, **1949**, *4a*, 321.

RECEIVED June 15, 1993

Chapter 4

Fluorescent Photoinduced Electron-Transfer Sensors

The Simple Logic and Its Extensions

Richard A. Bissell, A. Prasanna de Silva, H. Q. Nimal Gunaratne, P. L. Mark Lynch, Colin P. McCoy, Glenn E. M. Maguire, and K. R. A. Samankumara Sandanayake

School of Chemistry, Queen's University, Belfast BT9 5AG, Northern Ireland

The principle of photoinduced electron transfer is combined with the modular system 'Fluorophore-Spacer-Receptor' to develop the phenomenon of cation-responsive fluorescence. pH controlled 'on-off' fluorescence is demonstrated in the case of the dialkylaminoalkyl heterocyclic derivative **1a**. The modular system is then extended in two directions. In the first of these, targetting/anchoring modules are added to allow the investigation of proton fields in microheterogeneous membrane media with high spatial resolution. The sensor family **2a-f** is the realization of this approach. The second direction employs phosphorescent (instead of fluorescent) modules with/without protective shields to permit the development of phosphorescent pH sensing in an interference-free manner within intrinsically fluorescent environments. Sensors like **3** in aqueous β-cyclodextrin solution illustrate the possibilities of this idea.

Fluorescence is visual. Its appeal has much to do with the fact that it directly assaults our first sense *i.e.* vision. When fluorescence is expressed in molecules, we have a phenomenon of some power which is capable of bridging the divide between the world of molecules and that of our own with photons. Unlike many phenomena in chemistry, molecular fluorescence can be easily eliminated by chemical command *i.e.* quenching (*1,2*). Thus 'on' and 'off' states arise naturally. The possibility of negating or neutralizing the first command with a second results in fluorescence recovery and gives rise to reversible switching between 'on' and 'off' states.

Selective binding of a chemical species upon molecular recognition can lead to large perturbation of the host environment, especially when the guest is ionic. These perturbations can be exploited in a variety of ways in order to provide the first or second chemical commands that we require for fluorescence modulation. Conversely, the sensitivity of the host fluorescence to the guest occupancy endows the host molecules with chemosensory function. Our pet approach to such sensor molecules makes use of the photoinduced electron transfer (PET) principle, a subject which deservedly enjoys sustained popularity (3).

Fluorescent PET sensors can be formalized as 'fluorophore-spacer-receptor' systems of modular structure (4). The photon- and guest-interaction sites can be chosen to cater for various excitation/emission wavelengths and for various concentration ranges of a given guest (5). This choice has several constraints, however and these constitute the design logic of fluorescent PET sensors. The energy stored in the fluorophore excited state upon photon absorption must be sufficient to, say, oxidize the guest-free receptor and to simultaneously reduce the fluorophore (6). In essence, this is the thermodynamic criterion for an exergonic *trans*-spacer PET process and so far we and others have been blessed with fast PET rates which overwhelm the intrinsic emission of the fluorophore module. In effect, the first chemical command has been pre-programmed into the sensor system by building in the intramolecular receptor. If we consider a cationic guest it is easy to frustrate the PET process since the oxidation potential of the guest-occupied receptor is significantly higher than that of the guest-free receptor. The excited state energy of the optically pumped fluorophore thus remains unused and is returned as a photon (Figure 1). The cation entry is the second chemical command. Importantly, the spacer maintains the modularity of the system, which results in the additivity of component parameters (7,8), and is only violated by relatively long range forces. Fortunately for us, the PET process is of this type (9).

The fluorescent PET sensor design logic is flexible enough to give rise to a wide variety of examples and a selection from our laboratory is presented in Figure 2. Since most of these have received recent discussion (4,5), we now focus on a hitherto unpublished case **1a** which is representative of the general class of fluorescent PET sensors. The design of **1a** was directly inspired by the literature available for **1b** (14). Molecule **1b** possesses communication wavelengths in the visible region and also displays a distinctly lower fluorescence quantum yield than that of its parent **1c** which is devoid of the amino alkyl side chain. Sensor **1a** simply improves the electron donor ability of the side chain amine. While the unavailability of redox potential data for this heterocyclic fluorophore prevented an examination of the thermodynamic criterion for fluorescent PET sensor action of **1a**, the known behaviour of relatives **1b** and **1c** restored designer confidence. Indeed, strongly pH dependent fluorescence spectra are experimentally observed for **1a** in methanol:water (1:4, v/v), as expected for a good fluorescent PET sensor (Figure 3). The diethyl amino moiety efficiently quenches the fluorescence of the heterocycle across the dimethylene spacer until it is arrested by protonation at sufficiently low pH values (very low pH values of <2 cause quenching of fluorescence by direct protonation of the heterocycle in the excited state). 'On/off' function is clearly indicated by the large magnitude of the fluorescence enhancement factor of 100. The notable hypsochromicity observed in the electronic absorption spectra of **1a** in acidic media is caused by *trans*-spacer electric interactions between the protonated amine monopole and the dipole of the internal charge transfer (ICT) excited state (15) of the push-pull fluorophore. The wavelengths of maximum absorption under acidic and basic conditions, $\lambda_{\text{max.acid}}$ and $\lambda_{\text{max.base}}$ are 440 and 452nm respectively. We have recently described this type of behaviour for related cases (11). Linear regression analysis of these two sets of pH dependent spectra by means of equations (1) (10) and (2) (16) respectively, produces near-unity gradients and correlation coefficients.

$$\log [(I_{Fmax} - I_F)/(I_F - I_{Fmin})] = pH - pK_a \qquad (1)$$

$$\log [(OD_{max} - OD)/(OD - OD_{min})] = pH - pK_a \qquad (2)$$

It is notable that the ground state pK_a value is found in both the ground- and excited state experiments. The experimental values (± 0.1) are 8.7 and 8.5 respectively. This

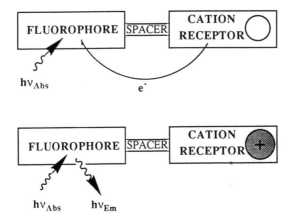

Figure 1. Diagrammatic representation of the design logic of fluorescent PET sensors for cations.

1a; R = CH$_2$CH$_2$N(C$_2$H$_5$)$_2$

1b; R = CH$_2$CH$_2$NH$_2$

1c; R = CH$_2$CH$_2$OH

1d; R = n-C$_4$H$_9$

1e; R = n-C$_8$H$_{17}$

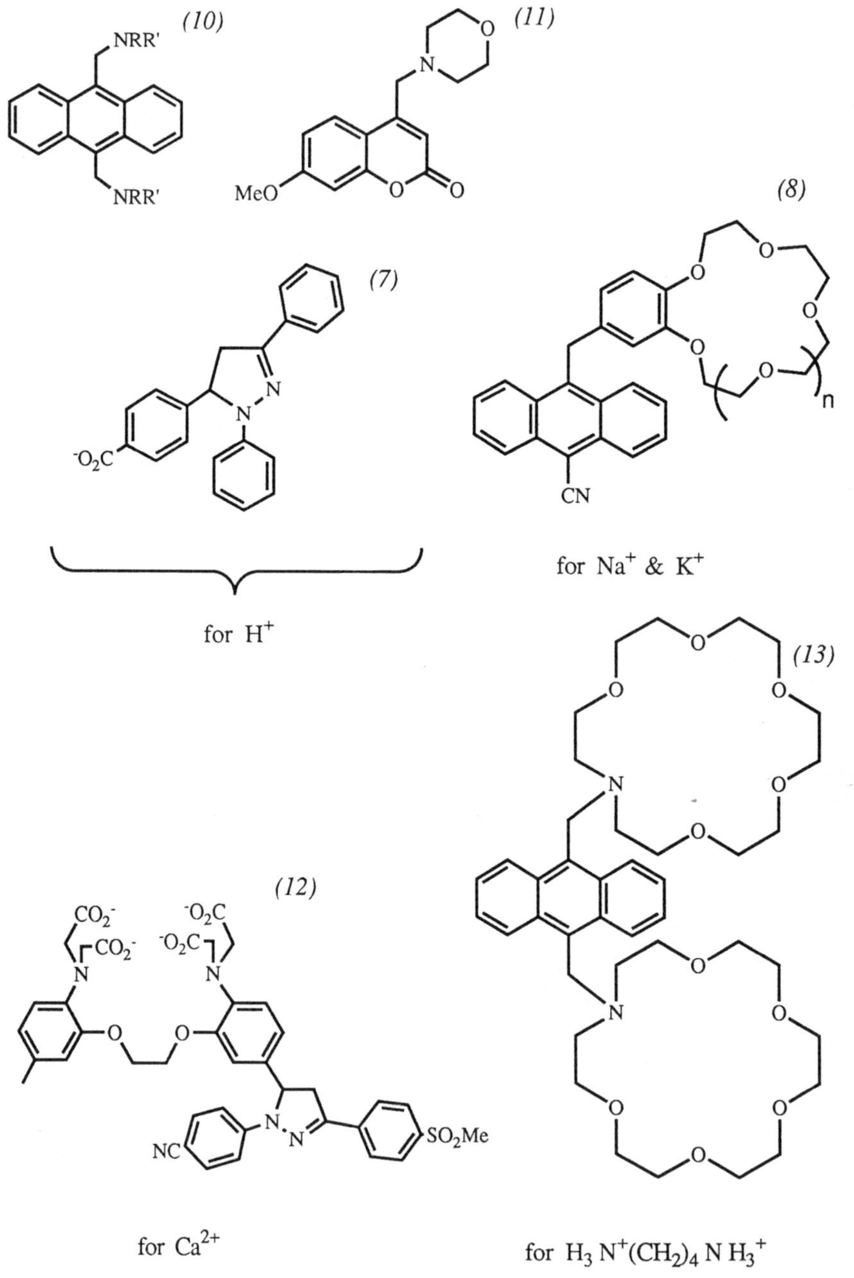

Figure 2. Previous examples of fluorescent PET sensors for various cations.

is a common feature of 'fluorophore-spacer-receptor' systems. In the present instance the *trans*-spacer electric interaction created by the ICT excited state must not be lasting long enough to re-equilibrate the protonation states of **1**. The fluorescence lifetime of **1e** is known to be 6.9ns in ethanol (*17*). Excited singlet state lifetimes of a few nanoseconds are found to be insufficient to establish new protonation equilibria even in cases where hindrances such as spacers are absent (*18*). In fact, analysis of the hypsochromicity according to equation (3) (*18*) suggests a theoretical pK_a (S₁) value of 7.5.

$$pK_a (S_1) - pK_a (S_0) = (hc/2.3RT)[(1/\lambda_{max.acid}) - (1/\lambda_{max.base})] \qquad (3)$$

It is also apparent that the fluorescence spectra for **1a** are blue shifted relative to that of the model parent fluorophore **1d**. The wavelengths of maximum emission are 499 and 516nm for **1a** and **1d** respectively. Since fluorescence is observed virtually exclusively from the monoprotonated form of **1a**, the origin of this emission hypsochromicity is easily assigned again to the ammonium monopole-excited ICT fluorophore dipole interaction. However, the difference in the present instance is that a thermally equilibrated ICT excited state must be involved here rather than the Franck-Condon ICT excited state encountered in the electronic absorption experiments. The monopole-dipole interaction energies in the two instances are 1.7 and 1.9 kcal mol⁻¹ respectively. Further, the experimental pK_a value differs from that of the parent receptor module triethylamine (11.0 in water) (*19*) due to the presence of an electron withdrawing moiety two methylene groups away. We can conclude therefore that the rich variety of heterocyclic fluorophores can be a valuable resource for fluorescent PET sensor designers, even though some aspects of the simple behaviour found with aromatic hydrocarbon fluorophores is degraded (*6*). Even with heterocyclic fluorophores the predictably simple behaviour can be recovered by the use of rigid, long spacers within the PET sensor format (*7*).

The strength of a design concept is considerably enhanced if expandability options are available. This is the case with fluorescent PET sensors and we present two lines of development here. First, we show how the addition of targetting and anchoring modules can permit us to use these molecular sensors to examine, and throw light on, problems in the nanometre domain. Second, we expand the fluorescent PET sensor concept beyond its apparent jurisdiction into another spin manifold - into phosphorescent PET sensor construction.

Mapping Membrane Bounded Protons with Molecular Versions of a Submarine Periscope

The molecular nature of fluorescent PET sensors can be put to good use when membrane bounded ions need to be monitored. For instance, membrane bounded protons lie at the heart of most energy transduction processes in biology (*20*). We consider detergent micelles since they are, in spite of their dynamic nature, the simplest and most convenient model membranes (*21*). Fluorescent pH sensors have been used in such contexts before (*22,23*), but the modular nature of our sensors allows an unprecedented approach to the mapping of proton concentrations near membrane interfaces. We specially note that fluorescent PET sensors are particularly useful for microenvironmental investigations since they have the detection sensitivity of excited state experiments (allowing low levels of sensor incorporation) as well as the thermodynamic certainty usually found only in ground state experiments. The latter point arises because fluorescent PET sensors consistently display ground state binding constants during excited state experiments as seen with **1a** for example. Figure 4 shows the extension of the three module 'fluorophore-spacer-receptor' supermolecule to include two targetting/ anchoring modules at the front and rear. Thus, we can anchor these sensors at a membrane-water interface with the proton receptor module protruding into the water to an extent which is controlled by the

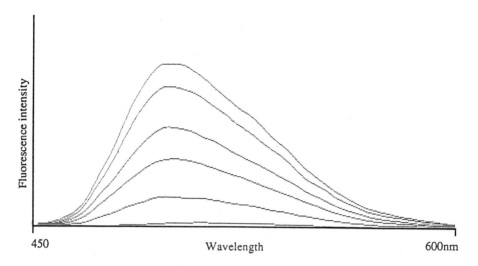

Figure 3. Family of uncorrected fluorescence emission spectra excited at the isosbestic wavelength 441nm for a 10^{-5}M solution of **1a** in water:methanol (4:1, v/v) with pH values (in order of decreasing intensity) of 4.7, 7.3, 8.2, 8.6, 9.1 and 11.3.

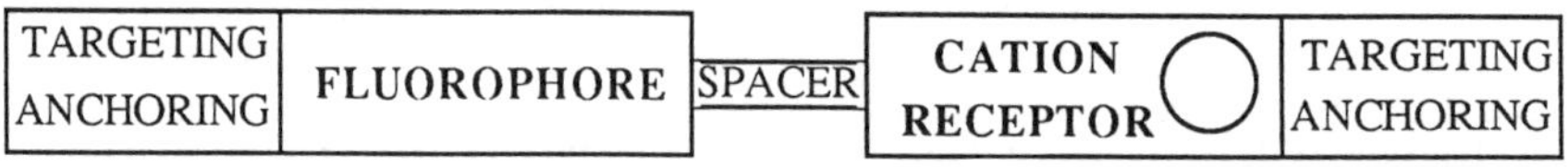

TARGETING ANCHORING	FLUOROPHORE	SPACER	CATION RECEPTOR ◯	TARGETING ANCHORING

Figure 4. Diagrammatic representation of a targettable fluorescent PET sensor.

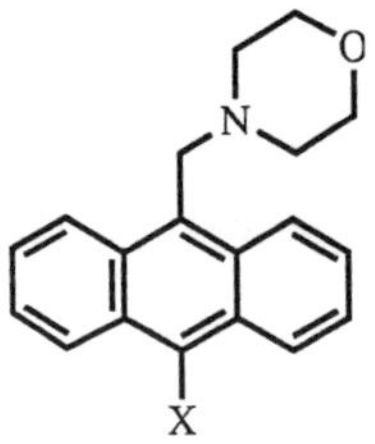

2a; X = n-$C_{18}H_{37}$

2b; X = n-C_8H_{17}

2c; X = Cl

2d; X = H

2e; X = SO_3^-

2f; X = $CH_2NH^+(CH_2CH_2)_2O$

hydrophobicity of the targetting head and tail groups. It is the position of the receptor module that concerns us most because it defines the proton catchment zone for the sensor. The other modules are only necessary for positioning the receptor and for transducing the local proton density into a humanly comprehensible optical signal. Such substitutional tuning of the microlocation of the receptor module for the purpose of depth (or height)-dependent measurement of effective proton density is reminiscent of the action of a submarine periscope for observations near a water surface.

Probes **2a-f** are such a family of molecular submarines which contain the tried and tested *(10)* 9-aminomethyl anthracene sensory core. In the cases of **2a-d**, the intrinsic hydrophobicity of the anthracene fluorophore would be expected to aid the anchoring of the sensor molecules among the polymethylene chains of the micelle monomer aggregate. **2a** & **2b** would have specially augmented hydrophobicities due to the 10-n-octadecyl or octyl chains. The tertiary amine, especially when protonated, is the least hydrophobic site in **2a-d** and this would be expected to orient towards the water phase. Cases **2e** & **2f** are the least hydrophobic sensors in this family and their orientations are less predictable. Fluorescence pH titration in various detergent solutions above their critical micellar concentration resulted in shape- and wavelength position-independent fluorescence spectra whose maximum intensities obeyed equation (1). The pH values are obtained with a macroscopic glass electrode and therefore represent values space averaged over all microenvironments with bulk water being the numerically dominant volume element. Essentially all of the features of the electronic absorption spectra are quantitatively conserved within experimental error. Before the resulting pK_a values (Table I) can be interpreted, it is crucial to note that the distal substituents will control the intrinsic basicity/acidity of the amino receptor unit across the anthracene short axis even though their intended primary purpose was for tuning the microlocation of the amine moiety. The intrinsic basicities of the various sensors **2a-f** were assessed by measuring their pK_a values in the non-micellar medium of methanol:water (1:4, v/v) with the results for **2a** & **2b** being estimated from that of **2d** for reasons of solubility and non-aggregation. These values are listed as $pK_{a.intrins}$ in Table I. ΔpK_a, the differential between an experimental pK_a value for a given sensor and its $pK_{a.intrins}$ gives us an estimate of the microenvironmental influence on the protons in the immediate vicinity of the proton receptor module of the sensor.

While the meaning of distance from the amine receptor to the micelle surface is degraded by its temporally dynamic and spatially convoluted nature, we can estimate the relative microlocation of a given member of a sensor series on a scale between the limits of bulk water and the less polar micelle interior in terms of the hydrophobicity constant (P) *(24)* of the variable component of the family **2a-f** *i.e.* the tail group. In other words, we can use hydrophobic free energy as the independent variable instead of a space coordinate for the mapping of membrane bounded proton densities. Such maps are best presented as free energy relationships *(25,26)* between ΔpK_a and log P values (Figure 5). These show clear periscopic effects and are discussed below.

If we first examine the profile for electrically neutral Triton X-100 micelles, we see that increasing distal group hydrophobicity causes more negative ΔpK_a values which soon reach a limit at -2.1. It is gratifying to find this smooth correlation for distal substituents of diverse charge types. This trend can be understood by the decreasing polarity of the microenvironment around the amine unit (as the sensor moves away from bulk water towards the micelle interface) being less able to solvate its protonated form. Thus, we see a serious dielectric constraint to protonation as we approach the micelle interface from the direction of bulk water. The profile for electrically positive CTAC micelles shows an augmented effect by reaching a limiting ΔpK_a value of -2.6 because of the additional electrostatic depletion of protons in the vicinity of the interface, *i.e.* the apparent basicity of the sensor is attenuated by both the local low polarity and the positive electric field strength. The outlier in this profile is **2e** which exhibits an apparent amplified hydrophobicity due to the

Table I. Acid/Base Characteristics of Sensors 2a-f in Micelle Media[a]

Sensor	$\log P_{C_6H_5X}$[b]	$pK_{a.intrins}$	$\Delta pK_a = pK_a - pK_{a.intrins}$		
			CTAC	Triton X-100	SLS
2a	+11.97[c]	6.3[e]	-2.6	-2.0	+2.2
2b	+6.57[c]	6.3[e]	-2.6	-2.1	+2.6
2c	+2.57	6.0	-2.6	-2.0	+2.4
2d	+2.10	6.3	-2.6	-1.6	+2.0
2e	-1.2[d]	6.6	-3.5	-1.1	+0.6
2f	-3.80[c]	5.1	-0.4	-0.2	+1.6

a. 10^{-6}M sensors were used in aerated solutions of water:methanol (4:1, v/v) (non-micellar medium), 5.0 x 10^{-3}M hexadecyl trimethylammonium chloride (CTAC, cationic micellar medium), 5.2 x 10^{-4}M polyoxyethylene (E 9-10) octyl phenol (Triton X-100, neutral micellar medium) and 2.0 x 10^{-1}M sodium dodecyl sulphate (SLS, anionic micellar medium) in water. The critical micelle concentrations of CTAC,Triton X-100 and SLS are 1.4 x 10^{-3}M, 2.6 x 10^{-4}M and 8.0 x 10^{-3}M respectively (22). The other terms are defined in the text.

b.   Logarithm of the partition coefficient of C_6H_5X between 1-octanol and water (24).

c.   Calculated value (24).

d.   Estimated from values in diethyl ether/water, i-butanol/water and relationships between these solvents and 1-octanol (24).

e.   Value for 2d used as mentioned in text.

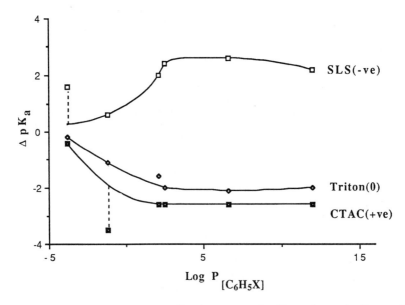

Figure 5. ΔpK_a - Log $P_{C_6H_5X}$ profiles for sensors 2a-f in various micellar media.

sulfonate(sensor) - trimethylalkyl ammonium(micelle) ion pairing. It is interesting that the ΔpK_a values of the sensors reach a roughly constant value of *ca*. zero irrespective of micelle surface charge type as the hydrophobicity of the targetting group becomes very low. The amine group is now measuring protons in the bulk water phase well away from the interface. As we move towards the electrically negative SLS micellar surface from bulk water, ΔpK_a values are influenced by two opposite effects. The micropolarity effect seen above is overcome by the influence of the negative electric field strength and the apparent basicity of the sensors are enhanced, *i.e.* the amine group encounters electrostatically concentrated protons. Sensor **2f** is the obvious outlier in this profile with another case of an apparent amplified hydrophobicity due to the protonated amine(sensor) - sulfate(micelle) ion pairing. Thus, in all three micellar media, the amine groups of the sensors **2a-f** are targetted to microlocations within the polarity and/or electric field strength gradient and these microlocations are a rational function of tail group hydrophobicity even though they hardly change above a certain hydrophobicity. The more general conclusion is that fluorescent PET sensors can be targetted to, and anchored in, gradually adjustable locations in microheterogeneous fields in order to transmit local information to a remote human observer.

Phosphorescent PET Sensing of Protons with a Molecular Message in a Bottle

While fluorescent sensing is a powerful idea, the incorporation of the sister phenomenon of phosphorescence (27) within the same conceptual framework would strengthen the PET sensing principle further. Also, phosphorescent sensing would open up that enticing possibility of tracking chemical species without any interference from the ubiquitous intrinsic/native fluorescence and scattering phenomena in living systems. Phosphorescence, being a delayed emission owing to the necessary spin conversion, can be easily time-resolved from fluorescence by applying a temporal gate of *ca*. 0.1 millisecond. This of course sets a limit to the time resolution but a whole host of interesting events happen over longer periods. Usually, the triplet excited states of phosphors are easily quenched by traces of molecular oxygen and other triplet states and therefore are difficult to observe in fluid solution at room temperature even after deaeration. Steric protection of the phosphor module to avoid material contact with its environment while allowing access to communication photons can be achieved by a transparent shield module. Figure 6 shows the extension of the 'lumophore-spacer-receptor' logic to introduce a transparent shield module around the lumophore (a generalized fluorophore). However, a regioselective self-assembly act is required here as the ion receptor must not be encapsulated if ion sensing is to remain viable. Cyclodextrins have been used for the past decade for enhancing the room temperature phosphorescence of organic guests in aqueous solution by encapsulation (28-35). The required regioselective insertion can be arranged by ensuring that the receptor module is less hydrophobic than the phosphor unit. In addition, the overall hydrophobicity of the three module molecule must be large enough and its three dimensional shape must fit into the cyclodextrin cavity in order to achieve sufficiently strong binding and a sufficiently slow exit rate constant. Under these conditions, the receptor module can protrude from the transparent shield to interact with and read from its microenvironment while the sterically protected lumophore translates this into a humanly readable message. Thus, we have a message in a bottle with an active cork and the message is read through the glass.

The molecular realization of the possibilities developed in the previous paragraph can be approached rationally. 1-Bromonaphthalene has been shown by Turro to be a strong phosphor at room temperature in deaerated β-cyclodextrin solution (28). Indeed, halonaphthalenes were among the examples employed by early workers to establish the favourable heavy atom effect on low temperature phosphorescence (36,37). The other essential piece of inspiration came from

Davidson's flash photolytic observation that the triplet excited state of 1-chloronaphthalene was quenched by tertiary amines in polar solvents (*38,39*). Thus we could reasonably expect that dialkylaminomethyl bromonaphthalenes such as **3** (*40*) in β-cyclodextrin hosts would have their phosphorescence switched off by intramolecular quenching. As mentioned in the previous paragraph, it is important that the dialkylamino group should be chosen such that it is less hydrophobic than the bromonaphthalene moiety. Also, the electron donor ability of the amino group must be kept favourable for PET processes. Protonation of the amine by a sufficiently acidic environment should then result in phosphorescence revival.

Sensor **3** in deaerated aqueous β-cyclodextrin solution does indeed show strong phosphorescence enhancement upon protonation (Figure 7). Application of a version of equation (1) to this data set yields a pK_a value of 6.9. Interestingly, this value is quite different to that obtained (7.9) by applying equation (2) to the pH dependent electronic absorption spectral data collected under the same conditions (except for deaeration). This difference is striking when we remember that the pK_a values obtained *via* absorption and emission experiments for **1a** were essentially identical. The crucial factor in the case of **3** is that here we are dealing with a triplet excited state which on average exists for 0.4 milliseconds. Such a time interval is more than sufficient to establish a new proton transfer equilibrium in the excited triplet state with significant internal charge transfer (ICT) character.

While the above results are most encouraging, they are incomplete until we show evidence for the molecular reality of the picture in Figure 6. The efficacy of the cyclodextrin in protecting the excited phosphor can be demonstrated by comparing the phosphorescence intensities arising from protonated **3** with and without β-cyclodextrin in water. β-Cyclodextrin causes a four-fold larger phosphorescence signal. While this factor can be larger for simpler phosphors like **4** (*28*), the cationic nature of protonated **3** permits the observation of some phosphorescence even in neat deaerated water. This suggests that bimolecular triplet - triplet annihilation is the dominant phosphorescence quenching channel under these conditions which is now being retarded electrostatically. Quenching due to interaction of the halonaphthalene lumophore with water is a less likely possibility, even though this mechanism dominates the photophysics of lanthanide complexes *via* the strong hydration of the central metal ion lumophore (*41*). Protrusion of the diethylamino moiety of **3** from the β-cyclodextrin host into the water can be experimentally supported by the fact that the pK_a value obtained from the pH dependent phosphorescence of **3** in water:methanol (9:1, v/v) is 7.2 which is within experimental error of the corresponding value seen in aqueous β-cyclodextrin solution.

In an effort to examine the scope of the 'message in a bottle' strategy and to develop phosphorescent pH sensors with longer excitation wavelengths, the case of **5** is examined. In contrast to sensor **3** which had a maximum extinction coefficient of 5300 $M^{-1}cm^{-1}$ at 283nm in methanol, **5** possesses the corresponding value of 15300 $M^{-1}cm^{-1}$ at 325nm. The quaternary salt **6** and other 1-bromo, 4-alkanoyl naphthalene derivatives were previously employed by Turro as phosphorescent probes of microheterogeneous systems (*42*) including β-cyclodextrin solutions (*29*) and provide the reason for the choice of **5** in our investigations. The pH dependence of the absorption spectra of **5** necessitates excitation at the isosbestic wavelength of 311nm for subsequent spectrophosphorimetry. Carbonyl compounds with less electron delocalization with nπ* triplet excited states can cause hydrogen atom abstraction from the cyclodextrin host (*43-45*) and can potentially interfere with the sensory action that we require. However, acetonaphthone derivatives possess ππ* excited triplet states (*46*) which are relatively ineffective for hydrogen abstraction and **5** is therefore not expected to show complications arising from irreversible

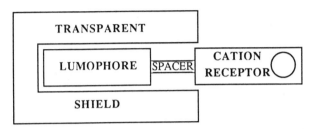

Figure 6. Diagrammatic representation of a sterically shielded phosphorescent PET sensor. Reprinted with permission from ref. 40. Copyright 1991 The Royal Society of Chemistry.

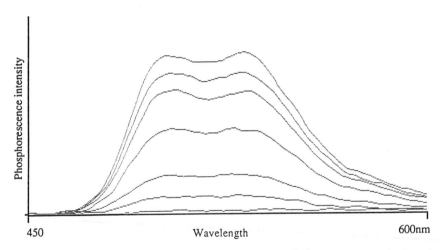

Figure 7. Family of uncorrected phosphorescence emission spectra excited at the isosbestic wavelength 283nm with an observation gate time of 4.0ms and a delay time of 0.1ms for a 5×10^{-5}M solution of **3** in deaerated water with 5×10^{-3}M β-cyclodextrin with pH values (in order of decreasing intensity) of 3.1, 5.6, 6.6, 6.9, 7.5, 8.0 and 9.0. Reprinted with permission from ref. 40. Copyright 1991 The Royal Society of Chemistry.

photochemistry. Indeed, **5** is found to be adequately photostable under our experimental conditions. Another possible irreversible photochemical pathway for both **3** and **5** could have arisen by debromination of the radical ion pair generated by the PET process which is central to the designed sensory mechanism. Again, this path is not found to detract from sensory behaviour, thanks to thermal back electron transfer which is expected to be rapid due to its intramolecular nature (*47,48*). Sensor **5** is not significantly complexed by β-cyclodextrin as inferred by the lack of any phosphorescence enhancement due to the macrocycle. This contrast with **3** can be attributed to the lower hydrophobicity of **5** with the carbonyl moiety as well as the observation that the phosphorescence of the 1,4 disubstituted isomer of **3** is also unaffected by β-cyclodextrin. Hydrophobic and geometric (size and shape) factors therefore apparently combine to disfavour **5** versus **3** for β-cyclodextrin encapsulation under phosphorimetric conditions. However, phosphorescence of **5** (λ_{max} 553nm) is observed to be strongly pH controlled in deaerated solution with a pK_a value of 6.0 and a hundredfold enhancement upon protonation. The success of **5** can be attributed, not to the strategy embodied in Figure 6, but to the charged nature of the emissive form of **5** as discussed in the previous paragraph concerning **3**. Thus in these instances, electric fields are in effect substituting for a molecular bottle. This is in some ways reminiscent of magnetic bottles in plasma physics where magnetic fields are employed for ion confinement. Overall, these results show that sterically or electrically protected phosphorescence can have a bright future in molecular sensing strategies for ionic species.

Conclusion

This article concentrated on just two of the several extensions to the fluorescent PET sensor design logic that are currently apparent. Most of these are outlined in a recent review of the research from this laboratory (*4*). There must be yet others over the horizon in uncharted waters, because the modular supermolecule approach is naturally expandable. Given the demonstrated flexibilty of the first generation systems (*4,5*) these conceptual extensions are likely to give rise to families of their own, with each member designed with a chosen set of optical and ion-binding properties in order to address specific problems. The future should be interesting for this avenue of supramolecular photophysics.

Acknowledgments

We are grateful to The Science and Engineering Research Council, Department of Education of Northern Ireland, The Queen's University of Belfast and The Nuffield Foundation for support. The help given by A. Morrow and Irene Campbell is very much appreciated, as are the thoughtful comments of a referee.

Literature cited

(1)   Guilbault, G.G. *Practical Fluorescence*, 2nd edn. Dekker: New York, 1990.
(2)   Lackowicz, J.R. *Principles of Fluorescence Spectroscopy*, Plenum: New York, 1983.
(3)   Fox, M.A. *Chem. Rev.* **1992**, *92*, 365.
(4)   Bissell, R.A.; de Silva, A.P.; Gunaratne, H.Q.N.; Lynch, P.L.M.; Maguire, G.E.M.; Sandanayake, K.R.A.S. *Chem. Soc. Rev.* **1992**, *21*, 187.
(5)   Bissell, R.A.; de Silva, A.P.; Gunaratne, H.Q.N.; Lynch, P.L.M.; McCoy, C.P.; Maguire, G.E.M.; Sandanayake, K.R.A.S. *Top. Curr. Chem.* **1993**, in press.
(6)   Bryan, A.J.; de Silva, A.P.; de Silva, S.A.; Rupasinghe, R.A.D.D. *Biosensors* **1989**, *4*, 169.
(7)   de Silva, A.P.; de Silva, S.A.; Dissanayake, A.S.; Sandanayake, K.R.A.S. *J. Chem. Soc. Chem. Commun.* **1989**, 1054.
(8)   de Silva, A.P.; Sandanayake, K.R.A.S. *J. Chem. Soc. Chem. Commun.* **1989**, 1183.
(9)   Wasielewski, M.R. In *Photoinduced Electron Transfer* ; Fox, M.A., Chanon, M. Eds.; part A. Elsevier: Amsterdam, 1988, p 161.
(10)   de Silva, A.P.; Rupasinghe, R.A.D.D. *J. Chem. Soc. Chem. Commun.* **1985**, 1669.
(11)   de Silva, A.P.; Gunaratne, H.Q.N.; Lynch, P.L.M.; Spence, G.L. **1993**, in preparation.
(12)   de Silva, A.P.; Gunaratne, H.Q.N. *J. Chem. Soc. Chem. Commun.* **1990**, 186.
(13)   de Silva, A.P.; Sandanayake, K.R.A.S. *Angew. Chem. Int. Ed. Engl.* **1990**, *29*, 1173.
(14)   Xuhong, Q.; Zhenghua, Z.; Kongchang, C. *Dyes and Pigments* **1989**, *11*, 13.
(15)   Shizuka, H. *Acc. Chem. Res.,* **1985**, *18*, 141.
(16)   Connors, K. *Binding Constants: The Measurement of Molecular Complex Stability*; Wiley: New York, 1987.
(17)   Alexiou, M.S.; Tychopoulos, V.; Ghorbanian, S.; Tyman, J.H.P.; Brown, R.G.; Brittain, P.I. *J. Chem. Soc. Perkin Trans. 2* **1990**, 837.
(18)   Ireland, J.F.; Wyatt, P.A.H. *Adv. Phys. Org. Chem.* **1976**, *12*, 131.
(19)   *Handbook of Chemistry and Physics* ; Weast, R.C. Ed.; 56th edn. CRC Press: Cleveland, 1975, p D-149.
(20)   Harold, F.M. *The Vital Force - A Study of Bioenergetics*; Freeman: New York, 1986.
(21)   Fendler, J.H. *Membrane Mimetic Chemistry*; Wiley: New York, 1982.
(22)   Kalyanasundaram, K. *Photochemistry in Microheterogeneous Systems*; Academic: Orlando, 1987.
(23)   Fendler, E.J.; Fendler, J.H. *Catalysis in Micellar and Macromolecular Systems*; Academic: New York, 1975.
(24)   Hansch, C.; Leo, A. *Substituent Constants for Correlation Analysis in Chemistry and Biology;* Wiley: New York, 1979.
(25)   *Advances in Linear Free Energy Relationships;* Chapman, N.B.; Shorter, J., Eds.; Plenum: London, 1972.
(26)   Kamlet, M.J.; Abboud, J.L.M.; Abraham, M.H.; Taft, R.W. *J. Org. Chem.* **1983**, *48*, 2877.

(27) Hurtubise, R.J. *Phosphorimetry: Theory, Instrumentation and Applications*; VCH: Cambridge, 1990.
(28) Turro, N.J.; Bolt, J.D.; Kuroda Y.; Tabushi, I. *Photochem. Photobiol.* **1982**, *35*, 69.
(29) Turro, N.J.; Okubo, T.; Chung, C. *J. Am. Chem. Soc.* **1982**, *104*, 1789.
(30) Turro, N.J.; Cox, G.S; Li, X. *Photochem. Photobiol.* **1983**, *37*, 149.
(31) Scypinski, S.; Cline Love, L.J. *Anal. Chem.* **1984**, *56*, 322.
(32) Scypinski, S.; Cline Love, L.J. *Anal. Chem.* **1984**, *56*, 331.
(33) Femia, R.A.; Cline Love, L.J. *J. Colloid Interfac. Sci.* **1985**, *108*, 271.
(34) Hamai, S. *J. Am. Chem. Soc.* **1989**, *111*, 3954.
(35) Munoz de la Pena, A.M.; Duran-Meras, I.; Salinas, F.; Warner, I.M.; Ndou, T.T. *Anal. Chim. Acta* **1991**, *255*, 351.
(36) McClure, D.S. *J. Chem. Phys.* **1949**, *17*, 905.
(37) Kasha, M. *Radiat. Res.*, **1960**, Suppl. *2*, 243.
(38) Beecroft, R.A.; Davidson, R.S.; Goodwin, D. *Tetrahedron Lett.* **1983**, *24*, 5673.
(39) Beecroft, R.A.; Davidson, R.S.; Goodwin, D.; Pratt, J.E.; Luo, X.J. *J. Chem. Soc., Faraday Trans. 2* **1986**, *82*, 2393.
(40) Bissell, R.A.; de Silva, A.P. *J. Chem. Soc. Chem. Commun.* **1991**, 1148.
(41) Horrocks, V.de W.; Sudnick, D.R. *Acc. Chem. Res.* **1981**, *14*, 384.
(42) Bolt, J.D.; Turro, N.J. *Photochem. Photobiol.* **1982**, *35*, 305.
(43) Monti, S.; Flamigni, L.; Martelli, A.; Bortolus, P. *J. Phys. Chem.* **1988**, *92*, 4447.
(44) Monti, S.; Cimaioni, N.; Bortolus, P. *Photochem. Photobiol.* **1991**, *54*, 577.
(45) Bohne, C.; Barra, M.; Boch, R.; Abuin, E.B.; Scaiano, J.C. *J. Photochem. Photobiol., A Chem.* **1992**, *65*, 249.
(46) Lim, E.C.; Li, Y.H.; Li, R. *J. Chem. Phys.* **1970**, *53*, 2443.
(47) Mauzerall, D.C. In *Photoinduced Electron Transfer*; Fox, M.A., Chanon, M. Eds.; part A. Elsevier: Amsterdam, 1988, p 228.
(48) Fox, M.A. *Adv. Photochem.* **1986**, *13*, 238.

RECEIVED April 27, 1993

Chapter 5

Tunable Fluorescence of Some Macrocyclic Anthracenophanes

H. Bouas-Laurent[1], J-P. Desvergne[1], F. Fages[1], and P. Marsau[2]

[1]Photophysique et Photochimie Moléculaire, Centre National de la Recherche Scientifique URA 348, and [2]Laboratoire de Cristallographie et de Physique Cristalline, Centre National de la Recherche Scientifique URA 144, Université Bordeaux 1, 33405 Talence Cedex, France

Based on the interplay between the complexing ability of crown-ethers or cryptands and the dual fluorescence (monomer/excimer) of aromatic rings (especially anthracene), new supramolecular receptors were designed and synthesized. Modulation of the fluorescence (wavelength, quantum yield, excimer/monomer ratio) of different photoactive systems is observed as a specific response to the complexation of metal (Na^+, K^+, Ag^+, Tl^+, Rb^+) or molecular (H_3N^+-$(CH_2)_n$-$^+NH_3$) cations. These properties could be used for light frequency conversion (anthraceno-coronand), detection and estimation of cations at low concentration (craterophanes, tonnelet) and evaluation of the solvent polarity (semaphorene).

Supramolecular chemistry is an actively developing interdisciplinary field with a great potential for a variety of applications *(1-5)*. Supramolecular systems, constructed from several molecular building blocks linked together by intermolecular or covalent bonds, show new properties or perform new functions that the components could not display separately. Particularly, the incorporation of a photoactive subunit can generate photoresponsive devices *(6)*. Some are based on the interplay between the complexing ability of crown-ethers or cryptands and the remarkable photophysical properties of the anthracene ring *(7-10)*.

In this chapter, we describe a series of supersystems (fluoroionophores) incorporating **one** or **two** anthracene rings, which undergo profound changes in fluorescence intensity and wavelength in the presence of protons and metals or molecular cations. The text is organized in the following way:

1. The anthracene ring as the fluorescing sub-unit
2. Anthraceno-coronands and Benzeno-coronands
3. Anthraceno-cryptands
   a. bicyclic receptors
   b. tricyclic receptors
4. Summary and Conclusion

0097–6156/93/0538–0059$06.00/0
© 1993 American Chemical Society

1. THE ANTHRACENE RING AS THE FLUORESCING UNIT

Anthracene and its derivatives are known as fluorophores. The fluorescence is excited at 360-400 nm and the emission is observed in the visible range (400-650 nm). Most often, the fluorescence is dual, i.e. the spectrum is modulated by the concentration and the temperature. An increase of concentration generates a new, red-shifted ($\bar{\nu}_o$-$\bar{\nu}_{Emax}$ ≈ 5000-7000 cm^{-1}) broad band (E) at the expense of the first emission (M) (see Figure 1). It was shown that this new emission is due to an **excimer**, a 1:1 complex, stable in the singlet excited state and dissociative in the ground state *(11-13)* according to the following equation (1):

$$M^* + M \rightleftharpoons E^* \tag{1}$$

$$\downarrow k_{FM} \qquad\qquad \downarrow k_{FE}$$

Intramolecular excimers are formed in bichromophores (equation 2) and the ratio of monomers over excimer fluorescence is controlled by the conformational mobility of the chain linking the two chromophores.

$$M^*\!\!\smile\!\!M \rightleftharpoons M^*M \tag{2}$$

$$\downarrow k_{FM} \qquad\qquad\quad \downarrow k_{FE}$$

The excimer stability depends on the degree of overlap between the π clouds of the two aromatic rings *(12)*; the best overlap results from a quasi-sandwich configuration *(14)*. Excimers constituted of two different chromophores, termed **exciplexes** have a stronger charge transfer character than excimers *(11-13)*; in this account, we consider intramolecular exciplexes formed from anthracene and amines.

The fluorescence quantum yield of anthracenes is generally high; internal conversion from the S$_1$ state of rigid polycyclic aromatic hydrocarbons is considered as negligible and the only competing channel is intersystem crossing to the T$_2$ state which is efficient enough for anthracene but can decrease sharply by substitution which lowers the S$_1$ energy level without affecting the T$_2$ energy a great deal *(15)*. Quantum yields for the fluorescing sub-units of interest in this chapter are given below (Figure 2).

A salient feature of anthracene among polycyclic aromatic hydrocarbons is the possibility to prepare a great number of derivatives in different positions; some of the derivatives can be obtained from commercially available substituted anthraquinones. This versatility is an advantage for the design of a variety of photoresponsive systems *(7-10 and 16-18)*.

2a. ANTHRACENO-CORONANDS

Taking advantage of the anthracene dual fluorescence, a cyclic system, **1**, incorporating two anthracene rings could be designed to detect the presence of cations by the formation of an intramolecular excimer as outlined in scheme 1. Such a system acts as a light frequency converter owing to the large spectral shift observed in the case of the following bis-anthraceno-coronand termed AAO$_5$O$_5$ (Fig. 3 and 4).

Single crystals of the free ligand display a typical monomer fluorescence in accordance with the absence of interaction between the rings whereas a typical

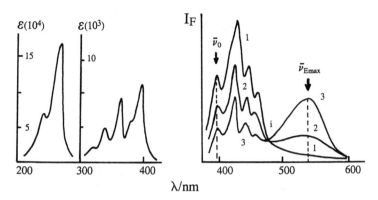

λ/nm

<u>Figure 1</u> Typical absorption and fluorescence spectra of an anthracene derivative. An excimer (E^*) is formed at the expense of the monomer (M^*) by increasing the concentration from 1 to 3 (v.g. from 10^{-5} to 10^{-2}M); (i) isolampsic point.

Z :   H nC_3H_7 OCH_3 —

ϕ_F :   0.30   0.76   0.48   0.80

<u>Figure 2</u>   Fluorescence quantum yields in degassed CH_3OH, of some anthracenes of interest in this account.

monomer (λM)   excimer (λE)

Fluorescence

<u>Scheme 1</u> Cation directed light frequency converter (see Fig.3 and 4).

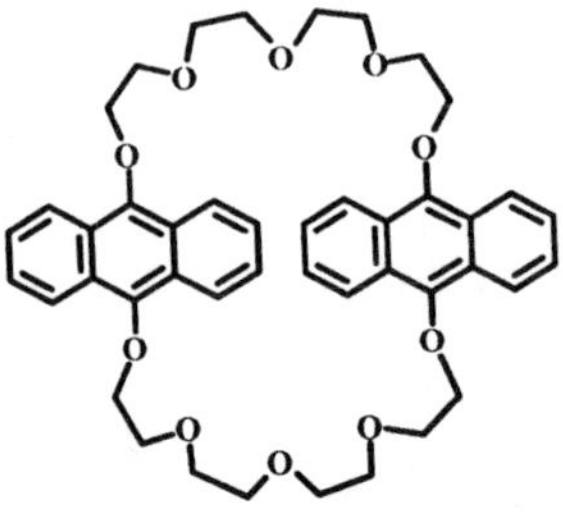

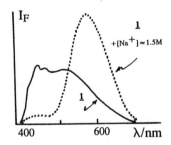

1 AAO₅O₅

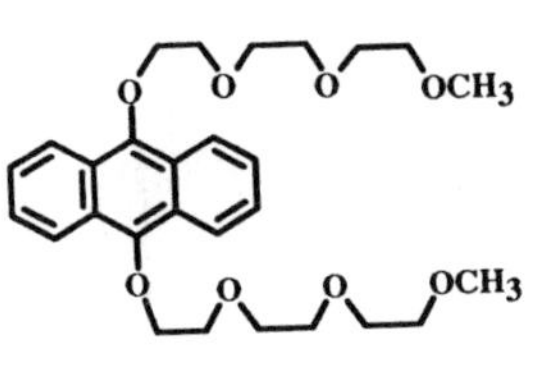

2

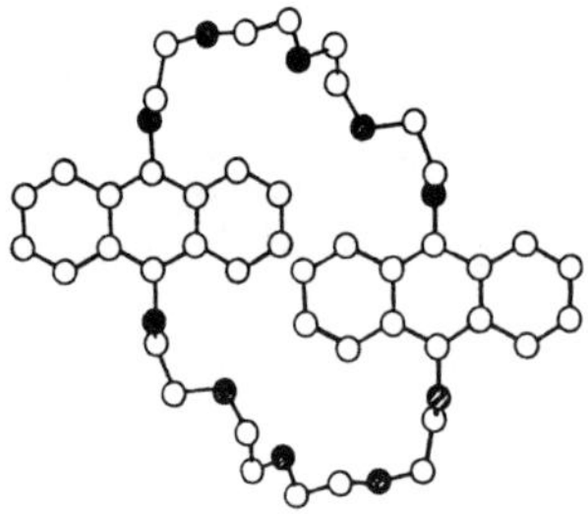

AAO₅O₅   2Na⁺,AAO₅O₅

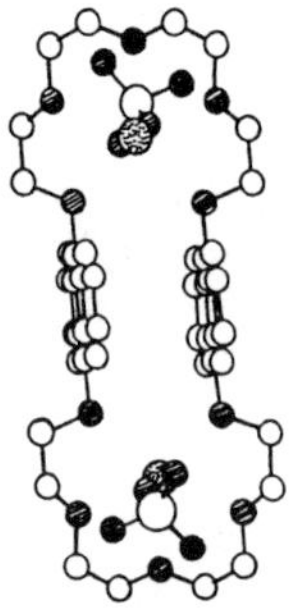

Fig. 3 Change of conformation by cation complexation of AAO₅O₅ (X-ray molecule structures reproduced with permission from ref. 9. Copyright 1991 VCH).

I_F

1

+[Na⁺]≈1.5M

1

400   600 λ/nm

Figure 4 Corrected fluorescence spectra of AAO₅O₅ in CH₃OH (——) free ligand conc ≈ 10⁻⁵M (•••••) in the presence of NaClO₄ in large excess. (reproduced from ref. 7. Copyright 1986 American Chemical Society).

excimer fluorescence (λ_{max} 580 nm) is exhibited by single crystals of the sodium complex $2Na^+ \subset AAO_5O_5$ *(7,19)* in which the two rings are in quasi sandwich arrangement (Fig. 3). An array of fluorescence quantum yields is observed in different solvents (Ether, CH_3CN, CH_3OH) for AAO_5O_5 but no wavelength shift. This is ascribable to the probable formation of some sort of a cluster between the coronand and the solvent leading to excimer quenching (Fig. 5).

The phenomenon is not shown by the reference molecule **2**. Recent work by others on anthracene derivatives are along these lines *(18-20)*.

2b. BENZENO-CORONANDS

Similar observations also apply to a related dissymetrical crown-benzenophane BBO_2O_5 (**3**) where the complexation of Na^+ induces an important modification of geometry reflected in the change of shape and maximum wavelengths of the fluorescence spectrum *(21)* (see Fig. 6a-6b). Compound **3** was shown by UV and fluorescence spectrometry as an hybrid between strained small and macrocyclic cyclophanes *(21,22)*.

3. ANTHRACENO-CRYPTANDS

Bicyclic Receptors.   Cryptands are known to form more specific and stable complexes with metal cation than coronands *(23)*; several mono anthraceno-cryptands were designed in order to encapsulate some cations in a given position *(8a)*. A consequence of the presence of two bridgehead tertiary amines in the proximity of the anthracene ring, is the easy formation of intramolecular exciplexes which, in polar solvents, usually lead to fluorescence quenching. This is the case of A_{33} (see formula Ann, n = 3).

Craterophanes (Ann).   9,10-anthraceno-cryptands (A_{22} and A_{33}) dubbed "craterophanes" because their structures suggest the shape of a vase were prepared to host metal cations above the central ring and at different distances from the ring.

The fluorescence of A_{33} in methanol is dual but the quantum yield is low ($\Phi_F \approx 0.04$) compared with that of 9,10-dipropyl anthracene, the reference compound ($\Phi_F = 0.76$). By complexation with an excess of metal cations such as K^+ ($\Phi_F = 0.30$) or protonation with an excess of CF_3COOH ($\Phi_F = 0.76$), a large fluorescence intensity increase is observed with the disappearance of the exciplex emission (Scheme 2); it is consistent with the fact that the nitrogen lone pairs interact no more with the aromatic ring but with the metal cations or the protons *(8a)*. It is thus possible to use this fluorescence enhancement to evaluate cations as shown in Fig. 7 for K^+ in CH_3OH. Other systems closely related to these cryptands designed and prepared by Czarnik *(20a)* and De Silva *(20b)* are described in this book.

Conversely, the fluorescence quantum yield of A_{22} in methanol is 0.68, a value close to that of the reference compound (0.76) and their spectra are superimposable; this suggests that in most of the A_{22} molecules, the nitrogen lone pairs are oriented out-out being hydrogen bonded with CH_3OH. Protonation or addition of K^+ or Na^+ does not affect the fluorescence very much but heavy metal cations such as Ag^+ and Tl^+ are very efficient quenchers *(8b)* (Scheme 3). X-ray structure determination of $Ag^+ \subset A_{22}$ and $Tl^+ \subset A_{22}$ (Fig. 8) has confirmed the 1:1 stoechiometry of the complexes and clearly shown that the cations sit inside the cavity, above the middle ring of the anthracene nucleus *(10,24)*. The η_6 coordination is a rather unusual situation for Ag^+ which is known to display η_2 hapticity *(25)*.

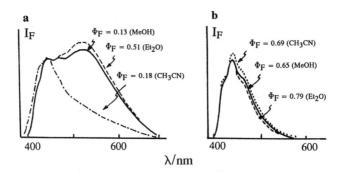

Fig. 5a Corrected fluorescence of $AAO_5O_5(\underline{1})$ at RT; conc $\approx 10^{-5}M$

Fig. 5b Corrected fluorescence of the reference molecule ($\underline{2}$) at RT; conc $\approx 10^{-5}M$

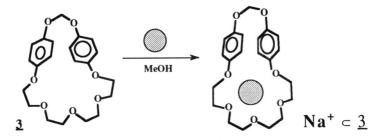

Fig. 6a Change of conformation of $\underline{3}$ by complexation with Na^+ (reproduced with permission from ref. 21. Copyright 1993 VCH).

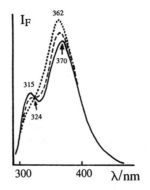

Fig. 6b Corrected fluorescence spectra in degassed CH_3OH of $\underline{3}$ ($10^{-5}M$),••••• without Na^+;− − −with Na^+,—— with Na^+ in large excess (taken from ref. 21, with permission).

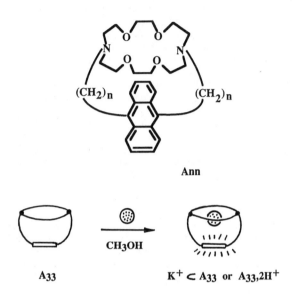

Ann

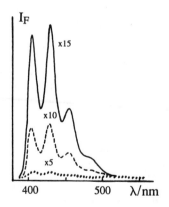

CH₃OH

A₃₃ $K^+ \subset A_{33}$ or $A_{33},2H^+$

Scheme 2 Enhancement of fluorescence of craterophane A₃₃. Complexation with a metal cation.

Figure 7 Fluorescence spectra of A₃₃ (10^{-5}M) in non degassed CH₃OH with increasing quantities of CF₃COOH in molar equivalents; •••• (0 to 5x) ; − − −(10x); —— (15x).

<u>Semaphorene</u>. Another cryptand of the same family, a 1,4-anthraceno-cryptand, was designed to extend the recently observed spectroscopic properties of 1,4-didecyloxyanthracene *(15)* which undergoes a strong solvatochromism in fluorescence emission and exhibits a high fluorescence quantum yield (Φ_F: 0.9 (methylcyclohexane), 0.98 (methanol) ...). Grafting a diazacrown ("N_2O_4") subunit to this anthracene derivative generated the new cryptand **4**, named "Semaphorene" for its shape and the intense visible light signals modulated by solvents *(26)* (see Fig. 9). Semaphorene fluorescence is also sensitive (λ_{max} and Φ_F) to the presence of a variety of cations, including Th^{4+} which behaves like H^+ *(26)*.

<u>Tricyclic Receptors: The "tonnelet"</u>. While crown-ether containing cyclophanes are being developed for the recognition of molecular guest cations *(4,27)* they are only few examples of hosts capable of detecting organic substrates by absorption and emission spectroscopy *(28)*. Bis-(9,10) anthracenyl macrotricyclic receptor **5** (Fig. 10a) combining the photophysical properties of the anthracene ring and the complexing ability of two face to face diazacrown macrocycles toward bis-alkylammonium cations, was prepared and shown to be the first example of selective optical detection of a linear molecular cation *(17)*.

Compound **5** was termed "tonnelet" (a French word meaning keg or cask) as suggested by inspection of the space filling molecular models. The tonnelet fluorescence spectrum (Φ_F: 0.085) is the envelope of the spectra displayed by the 9,10-dialcoxyanthracene ring (monomer type) and several intramolecular amine-anthracene and anthracene-anthracene excimer-type emission (Fig.11); this fluorescence was found to be very sensitive to the presence of α,ω-alkanediammonium salts, $Cl^-,H_3N^+-(CH_2)_n-^+NH_3$, Cl^-.

The cations that enter into the cavity of **5** (n > 5) form dihaptocryptates (Fig. 10b), the methylenic chain being held between the two aromatic rings (as confirmed by NMR) and an inhibition of the long wavelength emission is observed as a consequence, with an increase of the quantum yield (Fig. 11 and Table 1); the increase is particularly intense for n = 6 (Φ_F: 0.61) and the linear diammonium salts can be easily evaluated by this method (Fig. 12).

The binding stability constant were determined by UV absorption spectrometry (Table 1); it appears that the most tightly bound diammonium includes 7 methylene units and it does not correspond exactly with the best geometry for maximum fluorescence quantum yields; but, using fluorescence spectroscopy, the selectivity is in favor of n = 6 whereas the linear recognition is in favor of longer chain n $\geq$ 12 for the related 2,6-disubstituted macrotricyclic receptor **6** (Fig. 10-Table 1).

Examining the influence of a number of metal cations especially with K^+, Cs^+, Rb^+, Ba^{++} on the fluorescence spectrum of **5**, it was found that, **in contrast with the α,ω-alkanediammonium salts**, the spectra are broadened and displaced toward longer wavelengths (excimer-type) at the expense of the short wavelength component (monomer-type); the effect is particularly remarkable for Rb^+ (Fig. 13). That the long wavelength band peaking at 560 nm is that of an excimer is confirmed by comparison with the spectrum of the single crystal of 2 $Rb^+\subset$ **5**; the X-ray analysis provides evidence of the disalt formation (Fig. 14a) and clearly shows a quasi sandwich configuration of the two anthracene rings, in contrast with the structure of the free ligand where the rings prefer to lie remote from each other

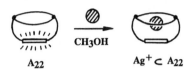

$$A_{22} \xrightarrow{\text{CH}_3\text{OH}} Ag^+ \subset A_{22}$$

Scheme 3   Fluorescence quenching by complexation in craterophane A_{22}

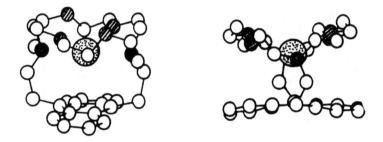

Figure 8   X-ray molecular structure projections along the anthracene ring long axis of $Ag^+ \subset A_{22}$ (left) and $Tl^+ \subset A_{22}$ (right); atomic sizes of Ag^+ and Tl^+ are not scaled (reproduced with permission from ref. 9. Copyright 1991 VCH).

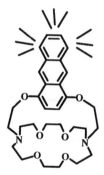

Semaphorene

Semaphorene **4**;   this "artist view" is close to the molecular structure determined by X-ray analysis (ref. 26).

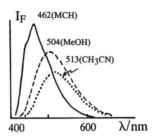

<u>Figure 9</u> Corrected fluorescence spectra of semaphorene in several solvents. $\Delta\tilde{\nu}$ (solvent-MCH): MeOH: 1800, CH_3CN: 2150 cm^{-1}.

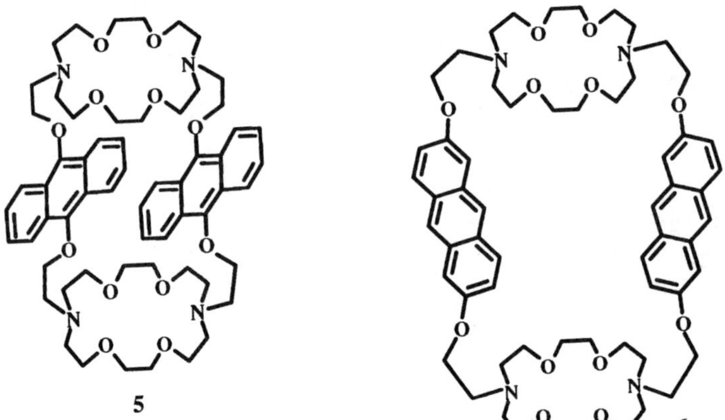

<u>Figure 10a</u> Idealised molecular formulas of the "tonnelet" **5** and its 2,6-isomer **6** (which experiences a longer distance between the two complexing sub-units).

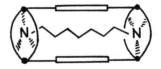

<u>Figure 10b</u>   An artistic view of a ditopic complex from α, ω - alkanediyldiammonium ions inserted in **5** or **6**.

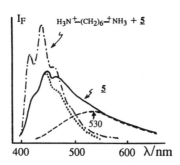

<u>Figure 11</u>   Corrected fluorescence emission spectra in degassed CH_3OH of $\underline{5}$ (——), Dimethoxyanthracene (DMOA) (•••), excimer-type (– – –) and in degassed CH_3OH: $CHCl_3$ 9:1 (v/v) of $\underline{5}$ ($10^{-5}M$) in the presence of H_3N^+-$(CH_2)_6^+NH_3$ ($10^{-3}M$) (adapted from ref. 17b. Copyright 1993 American Chemical Society).

<u>Table 1</u> - Quantum yields and stability constants (ß) of H_3^+N-$(CH_2)_n^+NH_3$ inclusion complexes with receptor $\underline{5}$ (CH_3OH: $CHCl_3$) 9:1 (V/V); 25°C; the selectivity is compared to that of $\underline{6}$ (reprinted from ref. 17b. Copyright 1993, American Chemical Society)

		$\underline{5}$		$\underline{6}$
n	Φ_F	log ß	$n=6$ Φ_F/Φ_F	$n=12$ Φ_F/Φ_F
5	0.11	-		-
6	0.61	6.78±0.5	1.0	-
7	0.55	7.16±0.5	0.9	0.39
8	0.42	6.04±0.5	0.69	0.58
9	0.51	5.20±0.5	0.84	0.84
10	0.51	5.20±0.5	0.84	0.95
12	0.21	4.60±0.7	0.34	1.0
tonnelet	0.085			

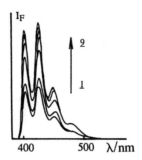

Figure 12   Fluorescence emission intensity of $\underline{5}$ (10^{-5}M) in non degassed CH$_3$OH: CHCl$_3$ 9:1 (v/v) as a function of added $^+$H$_3$N-CH$_2$)$_6$-NH$_3$$^+$ (25°C) conc. (10^{-5}M): $\underline{1}$: 0,0; $\underline{2}$: 0.25; $\underline{3}$: 0.50; $\underline{4}$: 0.75; $\underline{5}$: 1.0; $\underline{6}$: 1.5; $\underline{7}$: 2.0; $\underline{8}$: 2.5; $\underline{9}$: 3.0 (reprinted from ref. 17b. Copyright 1993 American Chemical Society).

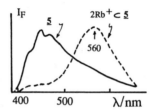

Figure 13   Corrected fluorescence spectra in degassed methanol (10^{-5}M, 25°C) of $\underline{5}$ (free ligand) and $\underline{5}$ in the presence of a large excess of Rb$^+$ClO$_4^-$ (Φ_{FE}/Φ_{FM}) is maximum for Rb$^+$ as compared with K$^+$ or Cs$^+$ (ref. 29).

(Fig. 14b) *(29)*. Interestingly, the stability constants for Rb$^+$ (log K$_1$ $\approx$ 5.8 ; log K$_2$ $\approx$ 4.8 in CH$_3$OH) does not indicate a cooperative effect *(29)*.

As a consequence of the above investigations, the tonnelet could be used for the selective optical detection of α,ω-diammonium salts (linear recognition) and Rb$^+$ (spherical recognition).

4. SUMMARY AND CONCLUSION

A variety of new supramolecular systems incorporating a complexing sub-unit (coronand-cryptand) and one or two photoactive sub-units (one or two anthracene rings) have been described. It was shown that the fluorescence emission is tunable according to the structure of the receptor and the nature of the medium and guest.

They can specifically and strongly affect the energy ($\bar{\nu}_{max}$) as well as intensity (Φ_F) of the emission and, in some cases, the excimer: monomer ratio (Fig. 15). As much as possible, the interplay between structure and fluorescence emission was studied, in parallel, in fluid solutions and in the crystalline state, which strengthens the structural interpretations in solution. Although the work described in this account was not purposely oriented toward analytical chemistry, some of the systems mentionned seem to have a great potential as sensors.

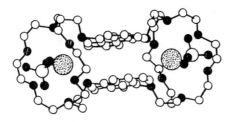

<u>Figure 14a</u> X-ray molecular structure of $\underline{2}$ Rb$^+ \subset \underline{5}$,projection perpendicular to the aromatic planes log K$_1$: 5.78 $\pm$ 0.08; log K$_2$: 4.8 $\pm$ 0.2 (reproduced with permission from ref. 17a. Copyright The Royal Society of Chemistry).

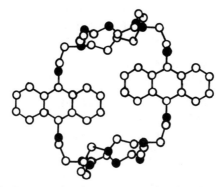

<u>Figure 14b</u> X-ray molecular structure of $\underline{5}$ (free ligand) (ref. 29).

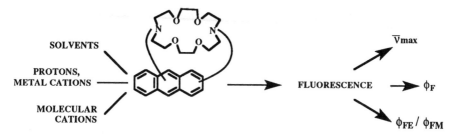

Figure 15 Factors affecting fluorescence emission.

Acknowledgements. This work was supported by the Centre National de la Recherche Scientifique (CNRS) and the Ministère de l'Education Nationale (MEN). The authors thank the colleagues and students whose names are cited in the references. We are particularly indebted to prof. J-M. Lehn, Dr A-M. Albrecht and prof. H. Hopf.

Literature Cited

(1)   Lehn, J-M., (a) Angew. Chem. Int. Ed. Engl., 1988, 27, 89; (b) *"Frontiers in Supramolecular Organic Chemistry and Photochemistry"*, Schneider H.J. and Dürr, H. Eds, VCH, Weinheim, 1991, 1-28.

(2)   Vögtle, F. *"Supramoleculare Chemie"*, Teubner, Stuttgart, 1989.

(3)   *"Supramolecular Chemistry"*, Balzani, V. and De Cola, L., Eds., NATO ASI series (II NATO Science Forum, Taormina, Sicily) Kluwer Acad., Dordrecht, 1992.

(4)   Gokel, G. *"Crownethers and Cryptands"*, Monographs in Supramolecular Chemistry, Stoddart J.F., Ed. The Royal Society of Chemistry, 1991.

(5)   (a) *"Supramolecular Photochemistry"*, Balzani V. Ed., Reidel, Dordrecht, 1987. (b) Balzani, V.; Scandola, F. *"Supramolecular Photochemistry"* Ellis Horwood, Kemp, T.J. Ed., New-York, 1991.

(6)   *"Fiber Optic Chemical Sensors and Biosensors"*, Wolfbeis O.S., Ed., CRC Press, Vol I and II, 1991.

(7)   Bouas-Laurent, H.; Castellan, A.; Daney, M.; Desvergne, J-P.; Guinand, G.; Marsau, P.; Riffaud, M-H. *J. Amer. Chem. Soc.* 1986, 108, 315.

(8)   (a) Konopelski, J-P.; Kotzyba-Hibert, F.; Lehn, J-M.; Desvergne, J-P.; Fages, F.; Castellan, A.; Bouas-Laurent, H. *J.C.S. Chem. Comm.* 1985, 433 (b) Fages, F.; Desvergne, J-P.; Bouas-Laurent, H.; Marsau, P.; Lehn, J-M.; Kotzyba-Hibert, F.; Albrecht-Gary, A-M.; Al-Joubbeh, M. *J. Amer. Chem. Soc.* 1989, 111, 8672.

(9)   Bouas-Laurent, H.; Desvergne, J-P.; Fages, F.; Marsau, P. in *"Frontiers in Supramolecular Organic Chemistry and Photochemistry"*, Schneider H.J. and Dürr H., Eds., VCH, Weinheim, 1991, 265-286.

(10)   Desvergne, J-P.; Fages, F.; Bouas-Laurent, H.; Marsau, P. *Pure and Applied Chem.*, 1992, 64, 1231.

(11)   Forster, Th. in *"The Exciplex"*, Gordon, M.; Ware, W.R. Eds., Acad. Press, New-York, 1975.

(12)   Birks, J.B. *"Photophysics of Aromaric Molecules"*, Wiley Interscience, New-York, 1970.

(13)   (a) *"Elements of Organic Photochemistry"*, Cowan, D.O.; Drisko, R.L. Plenum Press, New-York, 1976. (b) *"Essentials of Molecular Photochemistry"*, Gilbert, A.; Baggott, J. Blackwell, London, 1991.

(14)   Ferguson, J.; Morita, M.; Puza, M. *Chem. Phys. Letters* 1976, 42, 288.

(15)   Brotin, T.; Desvergne, J-P.; Fages, F.; Utermöhlen, R.; Bonneau, R.; Bouas-Laurent, H. *Photochem. Photobiol.* 1992, 55, 349.

(16)   Desvergne, J-P. and Bouas-Laurent, H., (a) *J.C.S. Chem. Commun.* 1978, 403. (b) *Israel J. Chem.* 1979, 18, 220.

(17)   (a) Fages, F.; Desvergne, J-P.; Bouas-Laurent, H.; Lehn, J-M.; Konopelski, J-P.; Marsau, P.; Barrans, Y. *J. Chem. Soc. Chem. Commun.*, 1990, 655. (b) Fages, F.; Desvergne, J-P.; Kampke, K.; Bouas-Laurent, H.; Lehn, J-M.; Meyer, M.; Albrecht-Gary, A-M. *J. Amer. Chem. Soc.* 1993, 115, 3658.

(18)   (a) Deng, G.; Sakaki, T.; Nakashima, K.; Shinkai, S. *Chem. Letters* 1992, 1287 (b) Deng, G.; Sakaki, T.; Kawahara, Y.; Shinkai, S. *Tet. Letters* 1992, 33, 2163.

(19)   (a) Guinand, G.; Marsau, P.; Bouas-Laurent, H.; Castellan, A.; Desvergne, J-P.; Lamotte, M. *Acta Cryst.* 1987, C43, 857. (b) Marsau, P.; Bouas-Laurent, H.; Desvergne, J-P.; Fages, F.; Lamotte, M.; Hinschberger, J. *Mol. Cryst. Liq. Cryst. Inc. Nonlin. Opt.* 1988, 156, 383.

(20)   (a) Huston, M.E.; Haider, K.W.; Czarnik, A.W. *J. Amer. Chem. Soc.* 1988, 110, 4460 (b) Prasanna de Silva, A.; de Silva, S.A. *J.Chem. Soc. Chem. Commun.* 1986, 1709.

(21)   Marsau, P.; Andrianatoandro, H.; Willms, T.; Desvergne, J-P.; Bouas-Laurent, H.; Hopf, H.; Utermöhlen, R. *Chem. Ber.* 1993, 126, 1441.

(22)   Hopf, H.; Utermöhlen, R.; Jones, P.G.; Desvergne, J-P.; Bouas-Laurent, H. *J. Org. Chem.* 1992, 57, 5509.

(23)   Lehn, J-M. *Angewandte Chem.* Int. Ed. Engl., 1988, 27, 89.

(24)   Marsau, P. et al., to be published.

(25)   Cohen-Addad, C.; Baret, P.; Chautemps, P.; Pierre, J-L. *Acta Cryst.* 1983, C39, 1343.

(26)   Desvergne, J-P.; Marsau, P.; Rau, J.; Cherkaoui, O.; Bouas-Laurent, H., to be published.

(27)   (a) Vögtle, F. "Cyclophan Chemie" Teubner, Stuttgart, 1990. (b) Diederich, F. *"Cyclophanes"*, Monographs in Supramolecular Chemistry, Stoddart J.F., Ed., The Royal Soc. of Chem. 1991. (c) Lehn, J-M., *Science* 1985, 227, 849. (d) Sutherland, I.O. *Chem. Soc. Rev.* 1986, 15, 63.

(28)   (a) Sutherland, I.O. *Pure Appl. Chem.* 1990, 62, 499. (b) Misumi, S. *Pure Appl. Chem.* 1990, 62, 493. (c) Chang, S.K.; Van Engen, D.; Fan, E.; Hamilton, A. *J. Amer. Chem. Soc.* 1991, 113, 7640. (d) Ueno, A.; Kuwabara, T.; Nakamura, A.; Toda, F. *Nature* 1992, 356, 136 and references therein.

(29)   Fages, F.; Marsau, P.; Desvergne, J-P.; Bouas-Laurent, H.; Barrans, Y.; Lehn, J-M.; Albrecht-Gary, A-M., to be published.

RECEIVED August 4, 1993

Chapter 6

Fluorescent Cyclodextrins for Detecting Organic Compounds with Molecular Recognition

Akihiko Ueno

Department of Bioengineering, Faculty of Bioscience and Biotechnology, Tokyo Institute of Technology, 4259 Nagatsuta, Midori-ku, Yokohama 227, Japan

Cyclodextrin derivatives with one or two chromophores were used as fluorescent sensors for detecting organic compounds in aqueous solution. They undergo induced-fit type of conformational changes associated with guest binding, changing the location of the chromophores mostly from inside to outside of the cavities. The guest binding abilities of the hosts are usually reflected in the sensitivities of the sensors, and various compounds including steroids were detected with remarkable molecular recognition.

There are many hosts with which metal cations can be detected by spectral changes. However, almost no effort has been invested to construction of the hosts that exhibit spectral variations upon complexation with guest molecules. Under this situation, we have attempted to construct host compounds that are spectroscopically responsive to molecules, and recently succeeded in constructing molecule-responsive sensors or indicators. This review gives a brief survey of our work.

Cyclodextrins (CDs) are doughnut shaped cyclic compounds consisting of six or more glucose units and form complexes of inclusion type with various organic compounds in aqueous solution (1). CDs are spectroscopically inert, but they can be converted into spectroscopically active compounds by modification with chromophores. We have prepared several chromophore-modified CDs in order to construct sensors or indicators for detecting molecules in aqueous solution, and found that they change fluorescence or absorption intensities on guest binding with different sensitivities depending on the shape, bulkiness, and polarity of the guest molecules.

0097–6156/93/0538–0074$06.00/0

Excimer Emission of Pyrene-modified γ-CD

CDs are named as α-, β-, and γ-CD for six, seven, and eight glucose members. α- and β-CD usually form 1:1 host-guest complexes (A) while γ-CD, which has a much larger cavity size, forms 1:2 host-guest complexes (B) in which two guest molecules are included in the γ-CD cavity (2,3). However, in the case of pyrene as a guest, γ-CD forms a 2:2 complex (C) because of the large size of pyrene (4). In this complex, two pyrene units are facing each other and thus form an excimer when one of pyrene units is excited. The excimer is an excited-state complex formed between singlet excited state and ground-state chromophores, and exhibits a broad fluorescence band in the longer wavelength region than that of normal fluorescence.

In connection with the 2:2 complex between γ-CD and pyrene, we prepared pyrene-modified γ-CD (D). This modified γ-CD forms a stable association dimer and exhibits a strong pyrene excimer emission around 470 nm (5). The association dimer was converted into a 1:1 host-guest complex upon guest addition (eq. 1 in Figure 1), and consequently the excimer emission decreased with increasing guest concentration (Figure 2). The extent of the decrease of the excimer emission intensity to the original one ($\Delta I/I°$) was used as a sensitivity parameter, and its values for various guests such as monoterpenes, steroids, alkaloids etc. were obtained. The results indicated that the sensitivity value is remarkably affected by size, shape, and polarity of the guest molecules. Thus, the pyrene-modified γ-CD was shown to act as a sensor for detecting organic compounds with molecular recognition (6). Among many guest compounds, the system was particularly sensitive to some steroids such as chenodeoxycholic acid and ursodeoxycholic acid. There was a good correlation between the fluorescence variations and the binding constants; therefore the strength of guest binding of the system is reflected in the fluorescence variation for each guest.

Sensors for Detecting Organic Compounds by Dansyl Fluorescence

Dansyl is known as a hydrophobic probe which exhibits fluorescence more strongly in a hydrophobic environment than in polar water milieue. Dansyl is also known to show fluorescence enhancement when its internal motions are restricted (7). We prepared dansylglycine-modified β-CD (β-DG, Chart 1) as a sensor, and observed that the fluorescence intensity around 560 nm decreased upon guest addition (8). This result suggests that the host undergoes a conformational change associated with inclusion of a guest molecule (eq. 2 in Figure 3), changing the location of the dansyl moiety from inside to outside of the cavity. This implies that the dansyl moiety is excluded from the hydrophobic environment of the CD cavity to water environment when the guest molecule is accommodated in the cavity. Various guest species (Chart 2) caused different decreases of the dansyl fluorescence (Figure 4). Among steroidal compounds, the host is not sensitive to ketosteroids such as progesterone, corticosterone, cortisone, prednisolone, and hydrocortisone while remarkably sensitive to chenodeoxycholic acid (CDCA) and ursodeoxycholic acid (UDCA). It is interesting that deoxycholic acid (DCA) was not effectively detected by the host in spite of the fact that it is also regioisomeric with CDCA and UDCA with one hydroxyl group at C-12 of the steroidal framework in place of C-7 of CDCA and UDCA. *l*-Borneol was detected with a comparable sensitivity to that of CDCA, while other compounds such as menthol, fenchone, nerol, and cyclohexanol were detected with moderate sensitivities. The result demonstrates that β-DG has remarkable molecular recognition ability and is useful as a fluorescence sensor for detecting organic compounds in aqueous solution. We also prepared dansyl-L-leucine-modified β-CD (β-DL), and found that its molecular recognition abilities are similar to those of β-DG although its sensitivities are roughly twice those of β-DG (Figure 4).

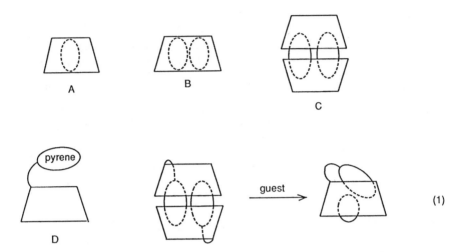

Figure 1. Complexes of CDs with different stoichiometries (A, B, C) and guest-induced dissociation of the dimer of pyrene-appended γ-CD (eq. 1).

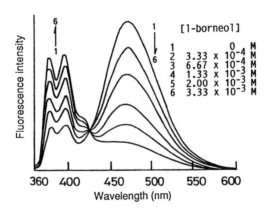

Figure 2. Fluorescence spectra of pyrene-appended γ-CD (6-deoxy-6-amino-[4-(pyrenyl)butanoyl]-γ-CD) at different *l*-borneol concentrations in 10% DMSO aqueous solution (Adapted from ref. 5).

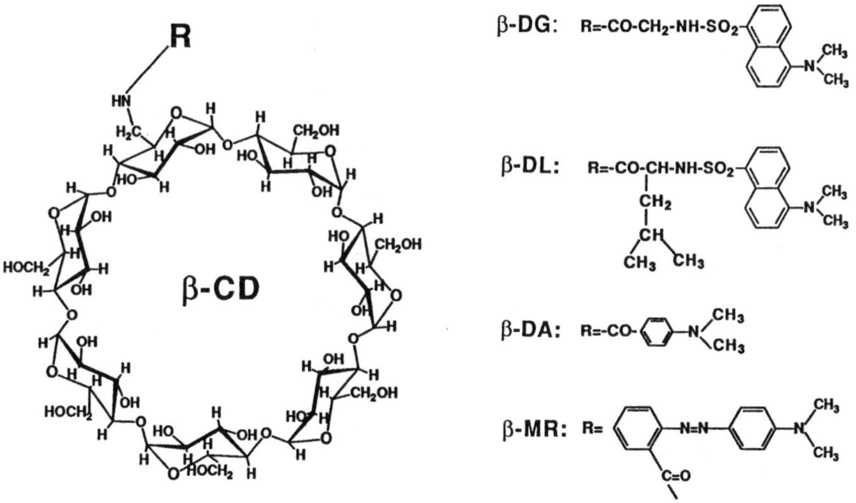

Chart 1.

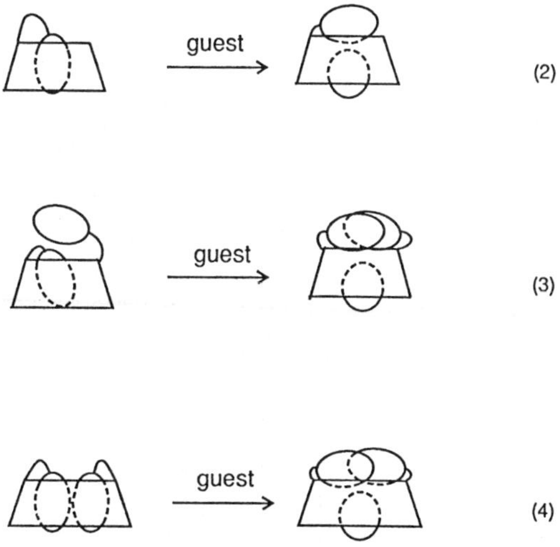

Figure 3. Induced-fit guest binding of mono-substituted (eq. 2) and di-substituted (eqs. 3 and 4) CDs.

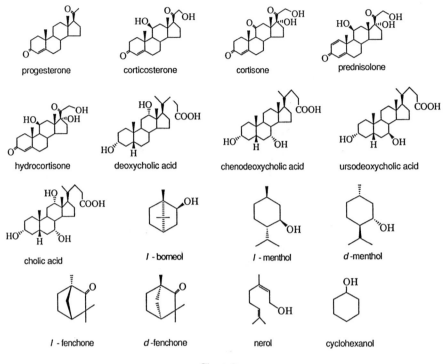

Chart 2.

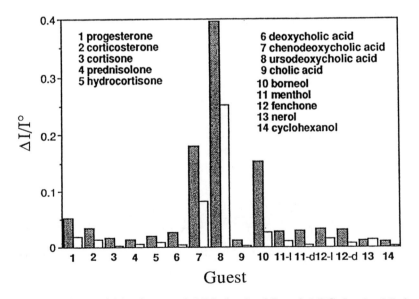

Figure 4. The sensitivity factors of β-DL ($\blacksquare$, 2 mM) and β-DG ($\square$, 2 mM) for various guests (10 mM) in aqueous solution.

Intramolecular Excimer Emission of Bichromophoric CDs

We prepared bichromophoric CDs as sensors which are capable of exhibiting excimer emission and changing its emission intensity upon guest binding. When glucose units of CDs are designated as A, B, C, D..... in the order along the CD ring, there are several regioisomers for bichromophoric CDs, three isomers of AB, AC, and AD for β-CD derivatives and four isomers of AB, AC, AD, and AE for γ-CD cases (Chart 3). We have prepared modified β-CDs (β-NS(A,X), X=B,C,D) and γ-CDs (γ-NS(AX), X=B,C,D,E), both bearing two naphthalene rings, and examined their fluorescence behavior. The β-CD derivatives includes one naphthalene in their cavities with the other locating outside the cavity, and consequently excimer formation is difficult. However, upon guest addition, they change the location of the one naphthalene from inside to outside of the cavity (eq. 3 in Figure 2), thus enabling mutual interaction between two naphthyl moieties outside the cavity (9). This conformational change is reflected in the guest-responsive fluorescence variation that excimer emission of AC and AD isomers is markedly enhanced upon guest addition (Figure 5). In the case of AB isomer, the enhancement in the excimer emission is slight, suggesting that the two naphthalene rings linked to A and B glucose units are difficult to take face-to-face orientation in this isomer in spite of the proximal location. On the other hand, γ-CD derivatives accommodate two naphthalene moieties and can form excimers except for the AB isomer, in which the face-to-face interaction is difficult similarly to the case of the AB isomer of β-CD derivatives. The excimer emission of the γ-CD derivatives are enhanced or depressed depending on guest species (10). The guest-induced excimer intensity variations of β-NS(AX) and γ-NS(AX) series are shown in Figure 6 for cholic acid, *l*-borneol, and cyclohexanol at 0.02 and 0.2 mM of host and guest concentrations, respectively (11). It is clear that the γ-CD derivatives such as the AD and AE isomers give large responses for the bulky guest, cholic acid, whereas the β-CD derivatives such as AD and AE give larger responses to *l*-borneol than those of the γ-CD derivatives. All of both series give negligible or very small responses to cyclohexanol, reflecting small binding constants of the guest for the hosts. The result demonstrates that a response pattern can be formed for each guest and the size of the guest is reflected in the pattern.

Excimer Emission of γ-CD Derivatives Bearing Two Pyrene Moieties

Recently, we have prepared a series of γ-CD derivatives bearing two pyrene moieties (γ-PC(AX), X=B,C,D,E), and examined the response patterns of the γ-CD derivatives (12). All of AB, AC, AD, and AE isomers exhibit predominant excimer emission around 520 nm, and show increase or decrease upon guest addition depending on the guest species except for the AB isomer that exhibits negligible fluorescence variations. It is interesting that *l*-borneol increases the excimer emission intensity while lithocholic acid decreases the intensity with concomitant enhancement in the normal fluorescence intensity around 420 nm (Figure 7). This result suggests that the structures of the complexes are different depending on the guest species. It is likely that the host compounds include two pyrene moieties in each cavity and the pyrene moieties in the complexes form highly hydrophobic environment around the guest species. Although the detail of the structures of the complexes is not yet clear, it is suggested that *l*-borneol is accommodated shallowly in the cavity, causing partial insertion of the two pyrene moieties into the cavity so as to make the excimer emission enhanced. On the other hand, it is suggested that lithocholic acid penetrates the CD cavity to exclude the two pyrene moieties from the cavity, thus decreasing the excimer emission intensity and increasing the monomer emission one. The binding constant of

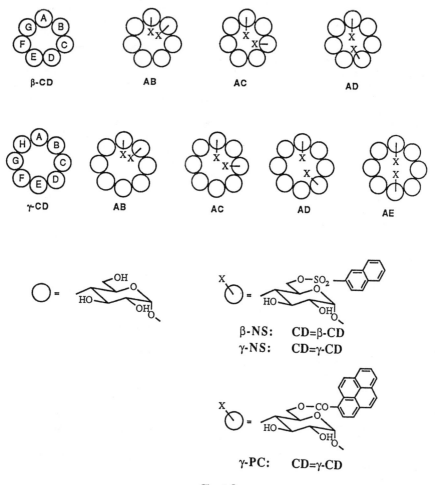

Chart 3.

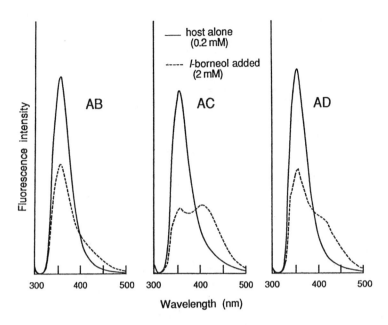

Figure 5. Fluorescence spectra of AB, AC, and AD isomers of β-NS series, alone (0.02 mM, ___) or in the presence of *l*-borneol (2.0 mM, ----) in 10% ethylene glycol aqueous solution (Reproduced with permission from reference 11).

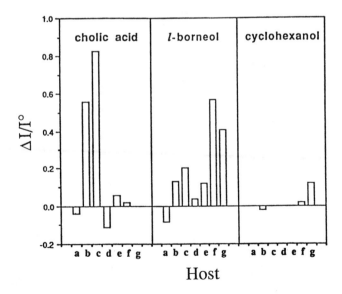

Figure 6. Guest-induced variations of excimer emission intensities of AB (a), AC (b), AD (c), and AE (d) isomers of γ-NS series and AB (e), AC (f), and AD (g) isomers of β-NS series. The concentrations of a-g and the guests are 0.02 and 0.2 mM, respectively (Reproduced with permission from reference 11).

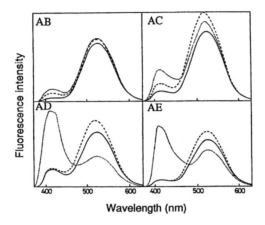

Figure 7. Fluorescence spectra of AB, AC, AD, and AE isomers of γ-PC series, alone (7.5 mM, —) or in the presence of *l*-borneol (1.0 mM, ---) or lithocholic acid (0.1 mM, ⋯⋯) in 30% DMSO aqueous solution (Reproduced with permission from reference 12).

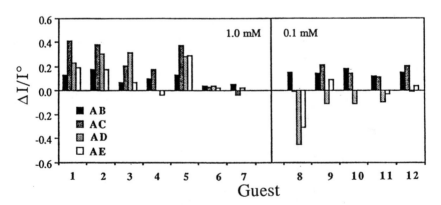

Figure 8. Guest-induced variations of excimer emission intensities of AB, AC, AD, and AE isomers of γ-PC series in 30% DMSO aqueous solution (7.5 mM). The guests are *d*-fenchone (**1**), *l*-fenchone (**2**), nerol (**3**), geraniol (**4**), *l*-borneol (**5**), *d*-menthol (**6**), *l*-menthol (**7**), lithocholic acid (**8**), deoxycholic acid (**9**), chenodeoxycholic acid (**10**), ursodeoxycholic acid (**11**), and cholic acid (**12**). The guest concentrations are 1.0 mM for **1-7** and 0.1 mM for **8-12** (Reproduced with permission from reference 12).

γ-PC(AC) are 19000 and 3700000 M^{-1} for *l*-borneol and lithocholic acid, respectively and the marked difference between the values seems consistent with the structural features described above.

Figure 8 shows response patterns of γ-PC(AX) series for various guests. Enantiomers of fenchone give similar response patterns, so that the degrees of asymmetric recognition of this series are low for fenchone. On the other hand, different response patterns were observed for a cis and trans pair of nerol and geraniol. In the case of steroids as guests, the AD isomer gave negative values while the AE isomer changes the sign of the response depending on steroids. These results indicate that the sensors are sensitive to the substituents of the steroid framework. The binding constants of this series are not consistent with the fluorescence variations, so that the guest-responsive behavior is suggested to be mostly determined by the structural factor of the complexes.

Sensors for detecting Organic Compounds by TICT Fluorescence

p-N,N-Dimethylaminobenzoic acid ester exhibits dual emission arising from non polar planar and twisted intramolecular charge transfer (TICT) excited states (13). Since the TICT emission is sensitive to solvent polarity, we prepared α-,β-, and γ-CD bearing a p-N,N-dimethylaminobenzoyl moiety as novel fluorescence sensors (14). Among these CD derivatives, β-CD derivative (β-DA) exhibits a strong TICT emission around 495 nm. The intensity of the TICT emission of β-DA decreased upon guest addition, and the system was found to be useful as a sensor for detecting various organic compounds.

A Color Change Indicator of Molecules

All above results have established that fluorescent CDs can be used as sensors of molecules. It is now desirable to construct color-change sensors or indicators of molecules on the same basis. From this viewpoint, we have prepared methyl red-modified β-CD (β-MR) as a guest-responsive color change indicator (15). Although methyl red itself changes its color from yellow to red around pH 4-5 with lowering pH, but the color of β-MR is still yellow even at pH 1.6 due to the fact that the methyl red (MR) moiety is included in the CD cavity and the protonation to the azo group is inhibited. However, β-MR changes its color to red upon guest addition. This result indicates that the MR moiety is excluded from inside to outside of the cavity associated with complexation of β-MR, thus being protonated in the acidic solution. Although there have been several reports on detecting metal ions by color changes, this is the first example of a color-change indicator for molecules.

Conclusion

Chromophore-modified CDs are useful for detecting molecules by fluorescence and absorption changes. The molecular recognition abilities of the chromophore-modified CDs are different depending on the chromophore and the kind of CD used. If we establish the methodology to design modified CDs capable of detecting a particular compound exclusively, any desired compound will be detected without the aid of biomolecules such as enzymes and antibodies.

Acknowledgment.

This research was supported by a Grant-in-Aid for Scientific Research from the Ministry of Education, Science and Culture of Japan.

Literature Cited.

(1) Cyclodextrin Chemistry; Bender, M. L; Komiyama M.; Springer-Verlag, Berlin, 1978
(2) Ueno, A; Takahashi, K; Osa, T. *J. Chem. Soc., Chem. Commun.* **1980,** 921.
(3) Ueno, A.; Moriwaki, F.; Osa, T.; Hamada, F.; Murai, K. *Tetrahedron* **1987,** 43, 1571.
(4) Herkstroeter, W G.; Martic, P. A.; Evans, T. R., Farid, S. *J. Am. Chem. Soc.* **1986,** 108, 3275.
(5) Ueno, A.; Suzuki, I.; Osa, T. *J. Am. Chem. Soc.* **1989,** 111, 6391.
(6) Ueno, A.; Suzuki, I.; Osa, T. *Anal. Chem.* **1990,** 2461, 1990.
(7) Kosower, E. M.; Huppext, D. *Ann. Rev. Phys. Chem.* **1986,** 37, 127.

(8) Ueno, A.; Minato, S.; Suzuki, I.; Fukushima, M.; Ohkubo, M.; Osa, T. *Chem. Lett.* **1990,** 605.
(9) Ueno, A.; Minato, S.; Osa, T. *Anal. Chem.,* **1992,** 64, 2562.
(10) Minato, S.; Osa, T.; Ueno, A. *J. Chem. Soc., Chem. Commun.* **1991,** 107.
(11) Minato, S.; Osa, T.; Morita, M.; Nakamura, A.; Ikeda, H.; Toda, F.; Ueno, A. *Photochem. Photobiol.* **1991,** 54, 593.
(12) Suzuki, I.; Ohkubo, M.; Ueno, A.; Osa, T. *Chem. Lett.* **1992,** 269.
(13) Rettig, W. *Angew. Chem. Int. Ed. Engl.* **1986,** 25, 971.
(14) Hamasaki, K.; Ikeda, H.; Nakamura, A.; Ueno, A.; Toda, F.; Szuki, I.; Osa, T. *J. Am. Chem. Soc.,* in press.
(15) Ueno, A.; Kuwabara, T.; Nakamura, A.; Toda, F. *Nature,* **1992,** 356, 136.

RECEIVED June 22, 1993

Chapter 7

Intrinsic Chromophores and Fluorophores in Synthetic Molecular Receptors

Thomas W. Bell, Daniel L. Beckles, Peter J. Cragg, Jia Liu, James Maioriello, Andrew T. Papoulis, and Vincent J. Santora

Department of Chemistry, State University of New York, Stony Brook, NY 11794-3400

As an alternative to the conjugation of *extrinsic* chromophores and fluorophores to biomolecular binding sites, light absorbing and emitting moieties can be incorporated into the molecular structure of a designed receptor. The resulting *intrinsic* chromophores and fluorophores can respond directly to complexation of analytes, such as metal ions and organic molecules, maximizing the probability that binding will produce useful optical effects. Macrocyclic ligands (*torands*), consisting of fully fused rings defining the molecular perimeter, exhibit useful fluorescence effects upon protonation and complexation of metal cations. Another macrocycle (**5**) undergoes ion selective tautomerism, producing a dramatic color change. Structurally related *hexagonal lattice hosts* bind organic molecules, such as guanidine, benzamidine and creatinine. Conformational changes, hydrogen bonding and polarization effects cause useful optical changes in the intrinsic chromophores and fluorophores in these synthetic molecular receptors.

Two very different approaches can be used for optical measurement of analyte concentration via selective complexation. The biomolecular approach employs ready-made binding sites that often display exquisite specificity. Molecular recognition is accomplished biologically but the coupling between binding and signal generation is a problem of chemistry. Conjugation of chromophores and fluorophores to antibodies and enzymes generally proceeds with poor site selectivity, resulting in uncertain communication with the binding site. Here the optical sensing component is *extrinsic* to the binding site, as represented in cartoon form in Figure 1. The alternate approach described in this chapter is to combine the binding and sensing functions in a single molecular subunit. As shown in Figure 1, incorporation of chromophores and fluorophores into a designed binding site produces an *intrinsic* optical sensing moiety. This maximizes the probability that local electronic polarization, desolvation and

0097–6156/93/0538–0085$06.00/0

conformational changes produced by complexation of the analyte to the binding site will produce useful optical effects.

At Stony Brook we have developed several families of artificial receptors for ions and neutral molecules consisting of fused heterocyclic and carbocyclic rings. The resulting torands (1-8) and hexagonal lattice hosts (9,10) are relatively planar molecules containing well-defined cavities or clefts lined with ligand atoms or hydrogen-bonding groups. The conformational preorganization of these hosts causes tight binding of complementary guests, yet their relative planarity engenders rapid and reversible complexation. The heterocyclic structures of these artificial receptors not only provide binding sites and facilitate shape-selective synthesis, they also constitute intrinsic chromophores and fluorophores. Conjugation of pyridine, naphthyridine and pyrrole rings with other unsaturated functionalities produces extended π-systems with HOMO-LUMO gaps corresponding to low energy UV-visible absorptions. Relative rigidity yields an added bonus: many of these synthetic receptors are strongly fluorescent. This chapter summarizes the results of an initial survey of the absorption and emission characteristics of these receptors containing intrinsic chromophores and fluorophores.

Torands

The torands are a relatively new series of macrocyclic ligands consisting of fully-fused rings that define the molecular perimeter. Three types of torands have been synthesized to date and are shown in Figure 2. Torands 1 (1-3,5-8,11) and 2 (12) are tri-n-butyl and triaryltri-n-butyl derivatives of dodecahydrohexaazakekulene (13). "Expanded" torand 3 (4,5,8) contains three additional pyridine rings, forming an alternating series of 1,8-naphthyridine and pyridine units, whereas 4 (14) is an expanded porphyrin with alternating pyrrole and pyridine rings. All of these torands have been synthesized in several steps from 9-n-butyl-1,2,3,4,5,6,7,8-octahydroacridine N-oxide (15,16). The overall yields are not high, but the reactions are sufficiently simple to perform on large scale that gram quantities of each are obtainable. The butyl chains apparently render these receptors sufficiently soluble that complexation experiments can be performed in many solvents ranging in polarity from methanol to chloroform. Water-soluble torands have not yet been prepared, but the synthetic methods employed would tolerate certain polar substituents. Torands substituted with functionalized alkyl or aryl groups would make it possible to study complexation in water.

Polypyridine Torand 1. Complexes of alkali metal triflates and picrates have been prepared from the monotriflate salt of torand 1, which is isolated directly from the macrocyclization reaction mixture. As shown in Figure 3, neutralization of the triflate salt with alkali metal hydroxides or carbonates gives the alkali metal complexes. With triflate or picrate counterions, torand complexes and salts partition selectively into chloroform rather than water. Thus 1 can be shuttled back and forth between the triflate salt and various complexes simply by washing the chloroform solution with aqueous acid or base. The free ligand is prepared by reaction of the triflate salt with tetra-n-butylammonium hydroxide in butanol/acetonitrile. Combustion microanalysis showed that the Li, Na, K, Rb and Cs complexes all have 1:1 host/guest stiochiometry, despite the cavity/ion size mismatch at both ends of the series. The X-ray crystal

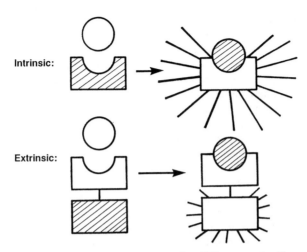

Figure 1. Cartoons illustrating potential differences in the optical responses of intrinsic and extrinsic chromophores and fluorophores in synthetic receptors.

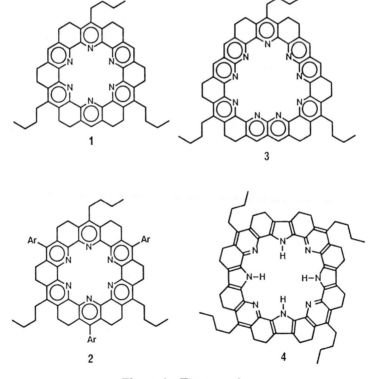

Figure 2. The torands.

structures of the Li, K and Rb picrate complexes (6,7) confirm that potassium is the best fit, since it is centered in the cavity (K-N 2.73-2.85 Å) and it lies only 0.23 Å from the best plane of the six nitrogen atoms. Rubidium lies 1.06 Å from this plane, whereas **1**•Li(picrate) crystallizes as a 2:2 complex in which each lithium atom is coordinated only to 2 or 3 N atoms and to two water molecules (one bridging and one terminal). Despite these structural differences, the Li, Na and K picrate complexes all have apparent stability constants in the narrow range of 10^{13}-10^{15} in water-saturated chloroform (3,11,12).

Torand **1** binds a wide range of metal salts. The following 1:1 complexes have been prepared in methanol and characterized by microanalysis: **1**•Ca(CF$_3$SO$_3$)$_2$, **1**•MnCl$_2$(H$_2$O)$_3$, **1**•Ni(NO$_3$)$_2$(H$_2$O)$_4$, **1**•Cu(NO$_3$)$_2$(H$_2$O)$_2$, **1**•CdCl$_2$(H$_2$O)$_4$, **1**•Cd(ClO$_4$)$_2$(H$_2$O)$_4$, **1**•Eu(NO$_3$)$_3$(H$_2$O)$_4$ and **1**•Gd(NO$_3$)$_3$(H$_2$O)$_3$(11, Cragg, P.J.; Kwok, D.-I., unpublished). Spectroscopic changes and precipitation of complexes demonstrated that **1** also binds Sr^{2+}, Ba^{2+}, Sc^{2+}, Ti^{4+}, Co^{2+}, Zn^{2+}, Ag$^+$, Sn^{4+}, Hg^{2+}, Tl$^+$, Pb^{2+}, Tb^{3+}, and UO$_2$$^{2+}$. Relatively subtle effects are observed in the UV-visible absorption spectra of **1**, its salts and metal complexes; representative absorption maxima are given in Table 1. The weak, longest wavelength absorption at 400 - 412 nm, apparently corresponding to an n→π* transition, varies little in wavelength or intensity. The major absorption at 340-350 nm also responds weakly to protonation or complexation. In contrast, protonation of **1** causes the appearance of a shoulder at 372 nm and an increase in the *fluorescence* intensity. Absorption and emission spectra are shown in Figure 4 for solutions of **1**•CF$_3$SO$_3$H in methanol. Certain metal ions, such as Pb^{2+} completely quench the fluorescence of **1** and this effect is under investigation for potential application to measurement of lead in solution.

The Dipyridotetrahydrocarbazole Chromophore; Expanded Porphyrin 4. In contrast with the polypyridine chromophore intrinsic to torand **1**, the dipyridotetrahydrocarbazole chromophore of torand **4** responds drastically to protonation (14). For example, Figure 5 shows the effect of adding HCl to an ethanolic solution of a diketone used in the synthesis of **4**. The absorption maximum shifts more than 90 nm from 364 to 456 nm, causing a visible color change from light yellow to deep amber. However, neither the free base diketone nor its HCl salt display any significant fluorescence in ethanol solution. The λ_{max} of the corresponding dipyridotetrahydrocarbazole lacking carbonyl groups also undergoes a bathochromic shift upon protonation (from 396 to 444 nm), but in this case the HCl salt is quite fluorescent, as shown in Figure 6. These two compounds are preorganized molecular clefts containing hydrogen-bond donors and acceptors. Hence, they are of interest as receptors for complexing organic compounds in their neutral forms and for binding anions in their polyprotonated forms. The absorption and emission effects displayed in Figures 5 and 6 apparently arise from differences in polarization, solvation and molecular conformation. Similar effects may be observed in complexation of receptors with intrinsic chromophores containing pyrrole rings, but this expectation awaits further investigation.

Expanded porphyrins are of current interest as ligands for multinuclear complexes and as cationic receptors for anions (17,18). Torand **4** (14) is the largest expanded porphyrin known and has potentially useful absorption and emission properties. Figure 7 shows the effects of protonation on the UV-visible spectrum of **4**. Solutions of free base **4** in ethanol give broad, indistinct absorbtions, suggesting

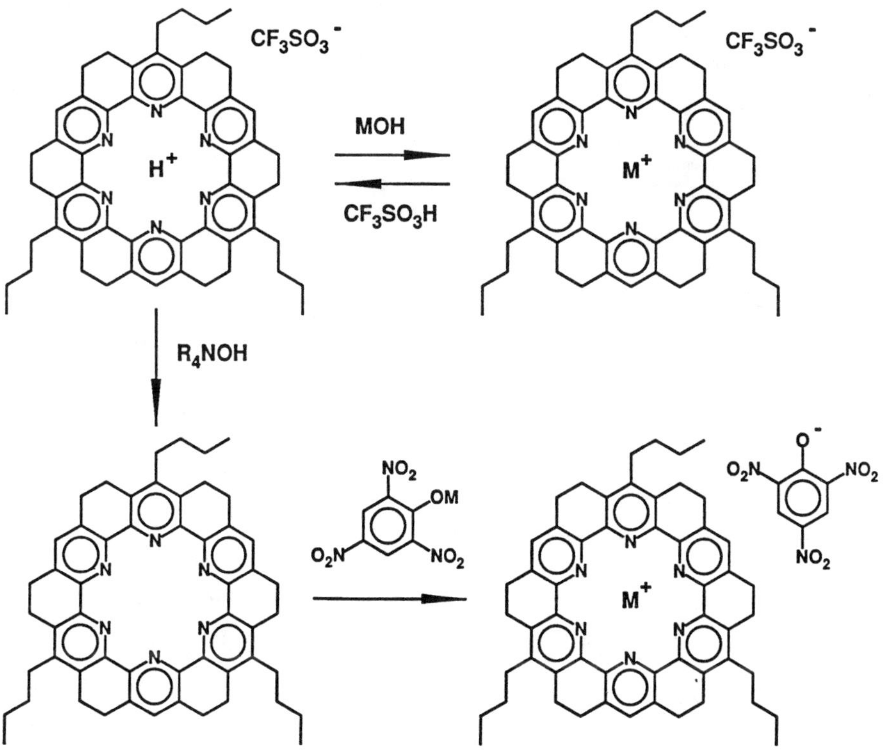

Figure 3. Synthesis of alkali metal triflate and picrate complexes and free ligand 1 from 1·CF$_3$SO$_3$H.

Table 1. UV-visible absorption maxima (λ_{max}) of methanol solutions of torand 1, its monotriflate salt and some metal complexes (br = broad; s = strong; sh = shoulder)

torand	412 (br)		344 (s)	
torand·HOTf	412 (br)	372 (sh)	340 (s)	
torand·ZnCl$_2$	400 (br)		350 (s)	
torand·MnCl$_2$	412 (br)		340 (s)	
torand·CoCl$_2$	408 (br)	388 (sh)	350 (s)	
torand·Ni(NO$_3$)$_2$	408 (br)		350 (s)	
torand·Eu(NO$_3$)$_3$	400 (br)	364 (s)	350 (sh)	318 (br)

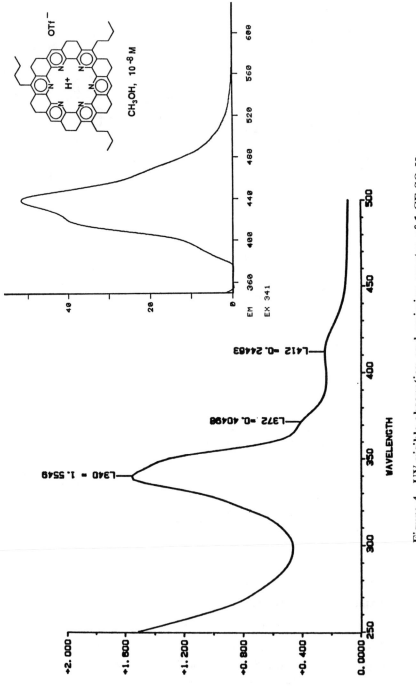

Figure 4. UV-visible absorption and emission spectra of **1·CF₃SO₃H** (CH₃OH, λ_{ex} 341 nm).

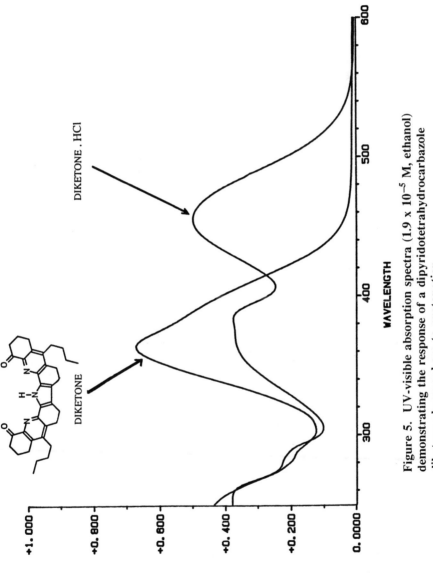

Figure 5. UV-visible absorption spectra $(1.9 \times 10^{-5}$ M, ethanol) demonstrating the response of a dipyridotetrahydrocarbazole diketone chromophore to protonation.

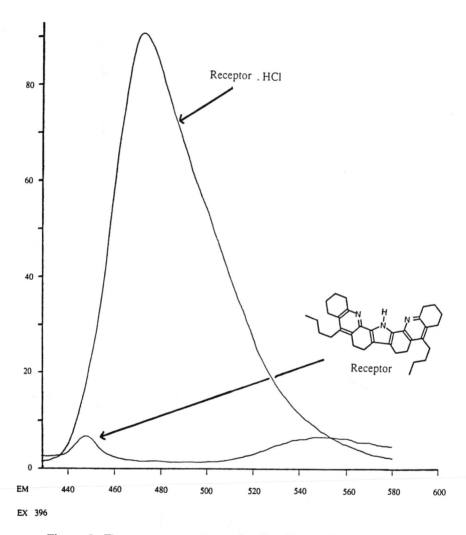

Figure 6. Fluorescence spectrum of a dipyridotetrahydrocarbazole chromophore (2.3 x 10^{-8} M, ethanol, λ_{ex} 396 nm) before and after addition of HCl.

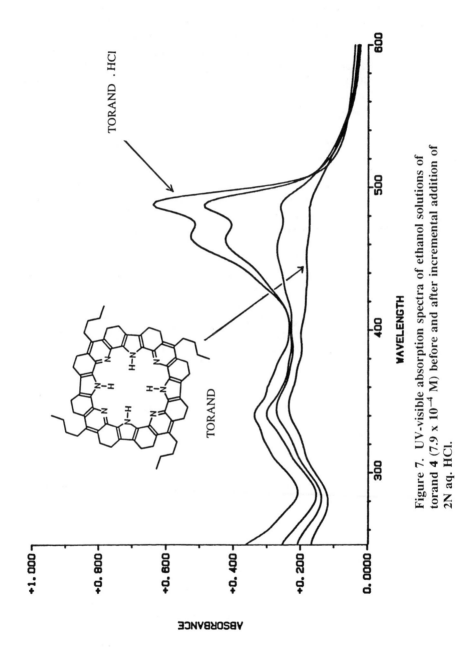

Figure 7. UV-visible absorption spectra of ethanol solutions of torand 4 (7.9 x 10^{-4} M) before and after incremental addition of 2N aq. HCl.

that a mixture of protonated and/or tautomeric forms may be present. Incremental addition of 2N HCl causes large increases in the intensities of the longest wavelength absorptions, producing deeply colored solutions. Solutions of free ligand **4** in ethanol are also fluorescent although the fluorescence is weaker than that observed for the dipyridotetrahydrocarbazole chromophore (Figure 6). The intensity of the emission band increases initially upon protonation, as shown in Figure 8, then decreases as a large excess of HCl is added. The observed fluorescence quenching may reflect anion complexation, but this speculation remains to be tested by recording the emission spectra in the presence of various anions.

Ion-Selective Tautomerism. In the course of synthetic studies related to expanded porphyrin torand **4**, we discovered a novel macrocycle that undergoes ion-selective tautomerism (*19*). Figure 9 shows that this 14-membered macrocycle (**5**) exists principally in the yellow-orange bis(hydrazone) form in aprotic solvents such as CH_2Cl_2 or $CHCl_3$. The bis(hydrazone) chromophore (λ_{max} 380 nm) is apparently stabilized by transannular hydrogen bonds. Complexation of a metal ion requires that these hydrogen bonds be broken and tautomerization to the nearly colorless bis(azine) form must occur. Yellow solutions of **5** in CH_2Cl_2 or $CDCl_3$ are decolorized when solid LiCl, $LiNO_3$, $Mg(NO_3)_2$ or $Ca(NO_3)_2$ are added, but weak optical responses are observed with other alkali metal and alkaline earth salts. The results show that **5** acts as an ion-selective ionophore that solubilizes small cations of high charge density with color change. It is not clear why the nitrate and chloride salts of individual metals respond differently, but this effect may be related to solubility differences in CH_2Cl_2. This ion-selective optical response of an intrinsic chromophore might be useful in the construction of reversible optical probes for analytical applications.

Hexagonal Lattice Hosts

Hexagonal lattice hosts are nonmacrocyclic receptors for organic molecules consisting of fused six-membered rings (*4,9,10*). The resulting molecular clefts are largely preorganized, but can undergo induced-fit complexation of guests by formation of hydrogen bonding networks. While torand **3** forms a strong complex with guanidinium chloride, the U-shaped receptor shown in Figure 10 binds guanidinium with conformational change (*4*). The UV-visible chromophore of **3** shows no significant response to complexation, whereas the U-shaped hexagonal lattice host binds guanidinium with bathochromic shifts of the longer wavelength absorption maxima (Figure 10). This receptor quantitatively extracts guanidinium chloride into anhydrous CH_2Cl_2 and titration studies were conducted for solutions in 95:5 ethanol/CH_2Cl_2. UV-visible absorption data recorded at 378 and 394 nm led to calculation of a stability constant of 2×10^6 L/mol for the 1:1 complex (*20*). Thus, guanidine complexation by this neutral receptor is relatively strong, yet reversible. Such receptors with intrinsic chromophores that have been tailored for maximum optical response could be used to construct selective chemical sensors.

A Benzamidine Receptor. Cationic aromatic guests may bind to synthetic receptors by a combination of hydrogen-bonding and π-stacking interactions. Figure 11 displays the complex of benzamidinium chloride with a hexagonal lattice cleft that is bridged by an aromatic spacer unit (*21*). Results of titration studies in 95:5

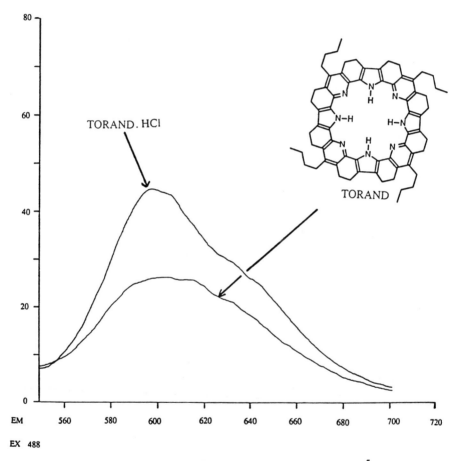

Figure 8. Fluorescence spectrum of torand **4** (1.6 x 10^{-5} M, ethanol, λ_{ex} 488 nm) before and after addition of HCl.

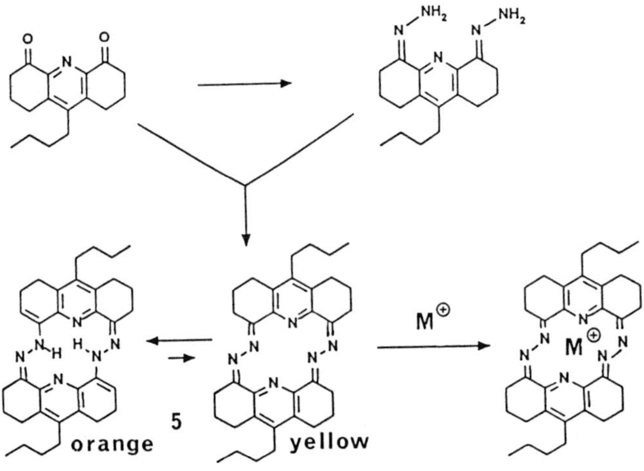

Figure 9. Bis(hydrazone)/bis(azine) tautomerism of a 14-membered macrocycle upon complexation of metal salts.

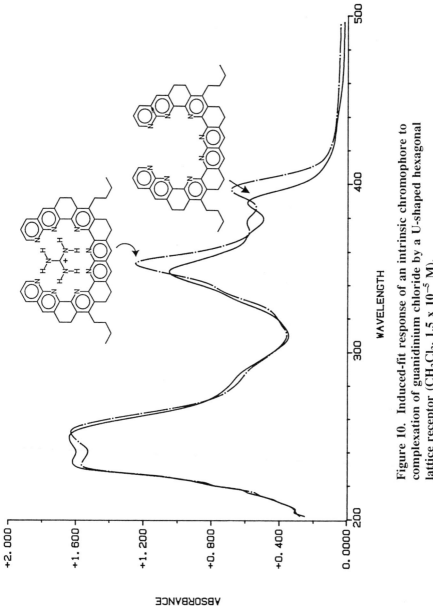

Figure 10. Induced-fit response of an intrinsic chromophore to complexation of guanidinium chloride by a U-shaped hexagonal lattice receptor (CH_2Cl_2, 1.5×10^{-5} M).

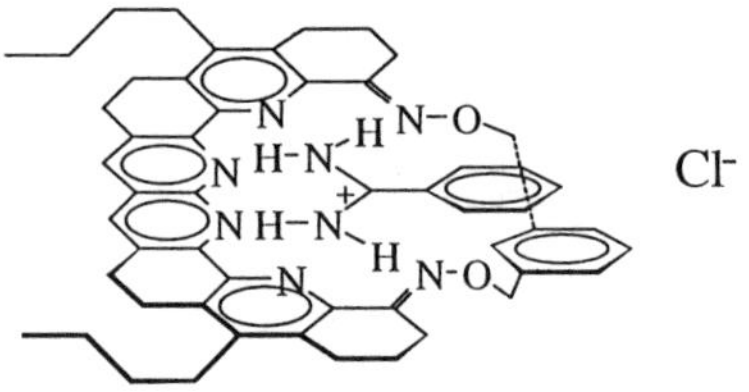

Figure 11. Hypothetical structure of the complex between a bridged hexagonal lattice cleft and benzamidinium chloride.

CH_2Cl_2/ethanol led to estimation of the stability constant of the 1:1 complex as 5 x 10^5 L/mol. Whereas the π-stacking interaction in complexes of modified receptors may produce an optical signal by electron transfer or tautomerization, the 6 nm bathochromic shift observed upon formation of the complex shown in Figure 11 apparently results from polarization or conformational changes in the heterocyclic intrinsic chromophore. A similar effect is observed in the corresponding anthracene-bridged receptor, as shown in Figure 12. To the extent that π-stacking stabilizes this complex the anthracene unit may be considered an intrinsic fluorophore. We anticipate that it will be possible to follow the complexation of benzamidinium salts by monitoring fluorescence changes and this approach is currently under investigation. Receptors for selective complexation of benzamidinium salts are of practical relevance to quantitation of pentamidine isethionate in biological fluids. This drug is used in the treatment of *Pneumocystis carinii* pneumonia, a leading cause of mortality in AIDS patients. Sensors and reagents based on this approach could prove superior to conventional chromatographic methods for analysis of therapeutic drugs (*22,23*).

A Creatinine Receptor. Urea and creatinine are blood metabolites of considerable importance in clinical diagnostic chemistry, particularly for evaluation of renal function (*24*). Earlier studies of urea complexation by a hexagonal lattice diketone have now been extended to binding of creatinine by synthetic receptors containing intrinsic chromophores and fluorophores (Figure 13, Beckles, D.; Maioriello, J.; Santora, V.; Chapoteau, E.; Czech, B. P.; Kumar, A., unpublished). In the urea complex six hydrogen-bond acceptors (**A**) converge on the binding site. Diagram **A** shows that the carbonyl oxygens and 1,8-naphthyridine nitrogens apparently form 4 hydrogen bonds, whereas the pyridines act as spacer groups and may stabilize the complex by dipole-dipole interactions. Creatinine exists in two tautomeric forms and may be complexed as the free base or as the conjugate acid. Diagrams **B**, **C** and **D** show three possible donor/acceptor arrays that may be incorporated into receptors for effective complexation of creatinine. The donor-acceptor array shown in **B** is preferred from the standpoint of secondary electrostatic interactions (*25*) and it was anticipated that one hydrogen-bond donor could be introduced by protonation of a pyridine ring under mildly acidic conditions.

An 1,8-naphthyridine o-aminonitrile moiety serves both as an effective donor-acceptor array for complexation of creatinine and as an intrinsic chromophore and fluorophore. In the pH range of 4.1-4.6 the monoprotonated form apparently predominates in 70 % aqueous methanol, producing the absorption spectrum shown in Figure 14. Under these conditions creatinine exists as a mixture of protonated and unprotonated forms, since its pK_a is approximately 4.2 in this solvent mixture. Such proton-transfer equilibria complicate the calculation of specific stability constants, but under buffered conditions absorption and emission changes result only from complexation, not from proton transfer. As shown in Figure 14, addition of creatinine to a buffered solution decreases the intensity of the 442 nm absorption band attributed to the protonated receptor. Creatinine complexation also quenches the yellow-green fluorescence of the protonated receptor and titration experiments in progress may yield the effective stability constant of the complex. This receptor exemplifies the manner in which intrinsic chromophores and fluorophores may be incorporated into hosts for reversible complexation of clinically important analytes (*26*).

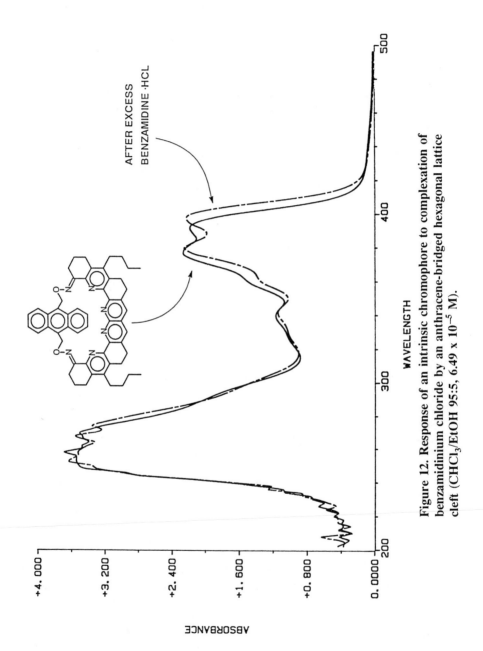

Figure 12. Response of an intrinsic chromophore to complexation of benzamidinium chloride by an anthracene-bridged hexagonal lattice cleft (CHCl$_3$/EtOH 95:5, 6.49 x 10^{-5} M).

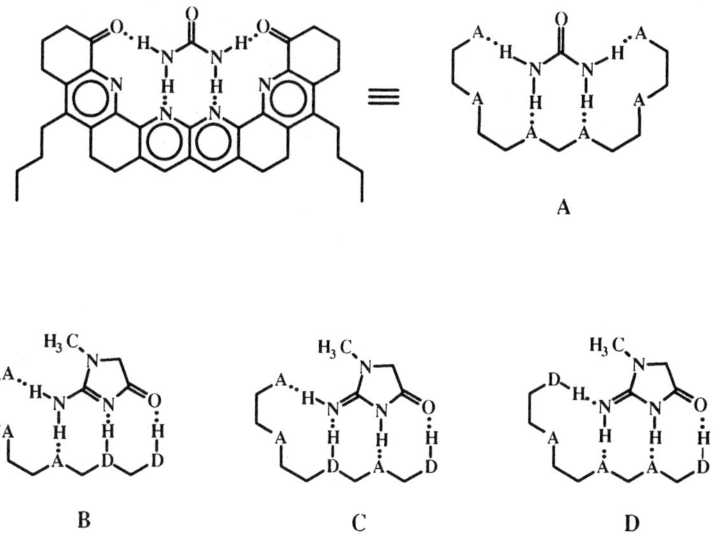

Figure 13. Receptor donor-acceptor arrays in a urea complex (**A**) and putative creatinine complexes (**B-D**).

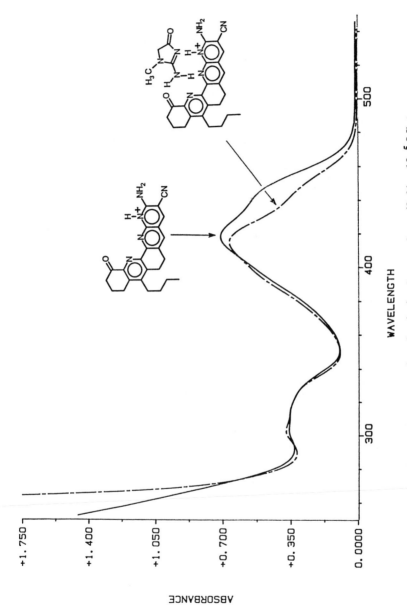

Figure 14. Response of an intrinsic chromophore (2.64×10^{-5} M) to complexation of creatinine in 70 % aq. methanol (buffered with MES, pH 4.1).

Acknowledgments. The National Institutes of Health (Grant GM 32937), the Center for Biotechnology at Stony Brook (funded by the New York Science and Technology Foundation) and Miles, Inc. are thanked for funding this research.

Literature Cited.
1. Bell, T. W.; Firestone, A. *J. Am. Chem. Soc.* **1986**, *108*, 8109-8111.
2. Bell, T. W.; Firestone, A.; Hu, L.-Y.; Guzzo, F. *J. Inclusion Phenom.* **1987**, *5*, 149-152.
3. Bell, T. W.; Firestone, A.; Ludwig, R. *J. Chem. Soc., Chem. Commun.* **1989**, 1902-1904.
4. Bell, T. W.; Liu, J. *Angew. Chem., Int. Ed. Engl.* **1990**, *29*, 923-925.
5. Bell, T. W.; Sahni, S. K. In *Inclusion Compounds Vol. 4: Key Organic Host Systems*; Atwood, J. E.; Davies, E.; MacNicol, D., Eds.; Oxford Univ. Press: Oxford, 1991; pp. 329-394.
6. Bell, T. W.; Cragg, P. J.; Drew, M. G. B.; Firestone, A.; Kwok, D.-I. A. *Angew. Chem., Int. Ed. Engl.* **1992**, *31*, 345-347.
7. Bell, T. W.; Cragg, P. J.; Drew, M. G. B.; Firestone, A.; Kwok, D.-I. A. *Angew. Chem., Int. Ed. Engl.* **1992**, *31*, 348-350.
8. Bell, T. W. In *Crown Compounds: Toward Future Applications*; Cooper, S. R., Ed.; VCH Publishers: New York, 1992; pp. 305-318.
9. Bell, T. W.; Liu, J. *J. Am. Chem. Soc.* **1988**, *110*, 3673-3674.
10. Bell, T. W.; Firestone, A.; Liu, J.; Ludwig, R.; Rothenberger, S. D. In *Inclusion Phenomena and Molecular Recognition*; Atwood, J. L., Ed.; Plenum: New York, 1990; pp. 49-56.
11. Firestone, A., Ph.D. Thesis, State University of New York, Stony Brook, NY, 1988.
12. Ludwig, R., Ph.D. Thesis, State University of New York, Stony Brook, NY, 1992.
13. Ransohoff, J. E. B.; Staab, H. A. *Tetrahedron Lett.* **1985**, *26*, 6179-6182.
14. Papoulis, A., Ph.D. Thesis, State University of New York, Stony Brook, NY, 1992.
15. Bell, T. W.; Rothenberger, S. D. *Tetrahedron Lett.* **1987**, *28*, 4817-4820.
16. Bell, T. W.; Cho, Y.-M.; Firestone, A.; Healy, K.; Liu, J.; Ludwig, R.; Rothenberger, S. D. *Organic Syntheses* **1990**, *69*, 226-237.
17. Sessler, J. L.; Cyr, M. J.; Burrell, A. K. *Synlett,* **1991**, 127-134.
18. Sessler, J. L.; Mody, T. D.; Ford, D. A.; Lynch, V. *Angew. Chem., Int. Ed. Engl.* **1992**, *31*, 452-455.
19. Bell, T. W.; Papoulis, A. T. *Angew. Chem., Int. Ed. Engl.* **1992**, *31*, 749-751.
20. Liu, J., Ph.D. Thesis, State University of New York, Stony Brook, NY, 1990.
21. Bell, T. W.; Santora, V. *J. Am. Chem. Soc.* **1992**, *114*, 8300-8302.
22. Lin, J. M.-H.; Shi, R. J.; Lin, E. T. *J. Liq. Chrom.* **1986**, *9*, 2035-2046.
23. Kulinski, R. F. *LC-GC* **1990**, *8*, 370-376.
24. Johnson, D. In *Clinical Chemistry*; Taylor, E. H., Ed.; Wiley: New York, 1989; pp. 55-82.
25. Jorgensen, W. L.; Pranata, J. *J. Am. Chem. Soc.* **1990**, *112*, 2008-2010.
26. U. S. patent application no. 07/705,733 filed March 28, 1991.

RECEIVED April 2, 1993

Chapter 8

Fluorescent Signal Transduction in Molecular Sensors and Dosimeters

Anthony W. Czarnik[1]

Department of Chemistry, Ohio State University, Columbus, OH 43210

Fluorophores equipped with binding groups serve to signal aqueous concentrations of metal ions, anions, and carbohydrates. The mechanism of signal transduction, photoinduced electron transfer, can be used to generate intensity-based readouts. A principle goal of our work has been to elucidate binding mechanisms that trigger concomitant changes in fluorescence.

In the past several decades, chemists have devoted their collective attention to the design of molecules that bind other molecules noncovalently. A primary goal of such work is the desire to engineer synthetic catalysts from first principles. However, an equally appealing goal, and one with enormous practical potential, is the design of chemosensors with selectivity for analytes such as metal ions (e.g., Pb^{+2}), anions (e.g., phosphate), and zwitterions (e.g., amino acids). The requisite research issues essential to the creation of such chemosensors are: (1) how can one bind a molecular entity with selectivity (preferably from water), (2) how can one generate signals from such binding events that are easy to measure, and (3) what mechanisms for binding and signal transduction intersect. Our group has focused on fluorescence as a signal transduction mechanism, both because of its potential sensitivity and because the multiplicity of mechanisms for affecting fluorescence provides for the incorporation of many known binding mechanisms. Furthermore, because fluorescence signalling is used in the creation of remote sensing applications with fiber optic techniques, there is a technological rationale for fundamental studies.

Fluorimetric methods have proven useful for the assay of metal ions in solution (1); for example, in vivo studies of calcium-selective fluorescence probes have been reported by Tsien (2). Most such analytical methods reported to date involve complexation of metal ions with aromatic heterocyclic ligands ("intrinsic" fluoroionophores). In 1977, Sousa described the synthesis of naphthalene-crown ether probes (Figure 1) in which the fluorophore π-system is insulated from the

[1]Current address: Parke–Davis Parmaceutical Research, 2800 Plymouth Road, Ann Arbor, MI 48105

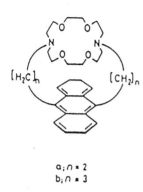

a; n = 2
b; n = 3

J. Chem. Soc., Chem. Commun. 1985, 433
Bouas-Laurent

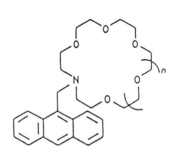

J. Chem. Soc., Chem. Commun. 1986, 1709
de Silva

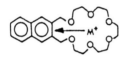

J. Am. Chem. Soc. 1977, *99*, 307
Sousa

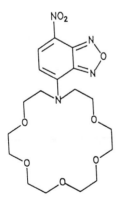

Anyl. Lett. 1986, *19*, 735
Street

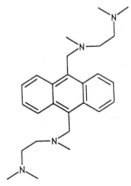

J. Am. Chem. Soc. 1988, *110*, 4460
Czarnik

Figure 1. Some fluorescent chemosensors for metal ions described by various authors.

donor atoms by at least one methylene group ("conjugate" fluoroionophores) (*3a*). These compounds demonstrated fluorescence changes upon the binding of alkali metal salts in an ethanol glass; the observed changes were attributed to a heavy atom effect (for Cs^+ and Rb^+) and/or a complexation induced change in triplet energy relative to ground and excited singlet state energies. Subsequent reports by various groups have built on this original premise, in which binding of metals to crowns and azacrowns has been coupled to emission changes of covalently attached fluorophores (Figure 1) (*3b-m*).

Chelation-Enhanced Fluorescence in 9,10-Bis(TMEDA)anthracene

Our work on the design of catalysts that associate <u>via</u> reversible covalent bond formation led us in 1987 to report the syntheses (Figure 2) of 9,10-bis(TMEDA)anthracene (**1**) and its bis($ZnCl_2$) chelate (**2**) (*4*). To our surprise, we found that the metal complex was over 1000-times more fluorescent than the free ligand (*3i*); the effect is readily apparent without the use of fluorimeter. A chelation-enhanced fluorescence (CHEF) effect of this magnitude proved unprecedented, and led us to consider both the nature of the fluorescence increase and extensions that might prove useful in providing a generalized "signal" to molecular recognition interactions.

Both the intensity and shape of compound **2**'s emission spectrum closely match that of 9,10-dimethylanthracene (Figure 3). What can account for this very large change in fluorescence emission intensity? We believe we are observing fluorescence quenching <u>via</u> exciplex formation, made very efficient in compound **1** by the high effective concentration of the intramolecular amine groups. Fluorescence quenching by inter- and intramolecular amines is, of course, a well-known phenomenon (*5*), and the fluorescence of 9,10-dimethylanthracene is quenched by addition of TMEDA as expected. At [DMA]=0.1 mM in acetonitrile, 4000 equivalents of added TMEDA result in quenching by a factor of 80; the intramolecular quenching at the same concentration of compound **1** with no added TMEDA is roughly 30-times more efficient than even this. Consequently, we explain the observed CHEF in 9,10-di(TMEDA)anthracene by noting that, when chelated to a metal ion, the amine lone pairs become involved in bonding and are unable to donate an electron to the excited state of the anthracene. This explanation is corroborated by the pH profile shown in Figure 4 obtained in aqueous solution. At pH 11.7, the amine groups are almost completely unprotonated and the fluorescence of compound **1** is consequently very low; at pH 1.6, the amines (most or all; one cannot distinguish based on the data in Figure 4) are protonated, and the fluorescence increases by a factor of over 300. In acetonitrile, the fluorescence increase may be titrated by addition of metal ion (Figure 5). Solubility limits prevented us from further raising the metal ion concentration and therefore determining an asymptotic "intrinsic fluorescence" for complex **2**, which is partially dissociated even in acetonitrile saturated with $ZnCl_2$. The same titration done in aqueous solution fails; it is not surprising that the complexation of zinc ion with **1** in water is not complete at equimolar concentrations, and we predicted that cryptands leading to polydentate chelation would provide much larger association constants.

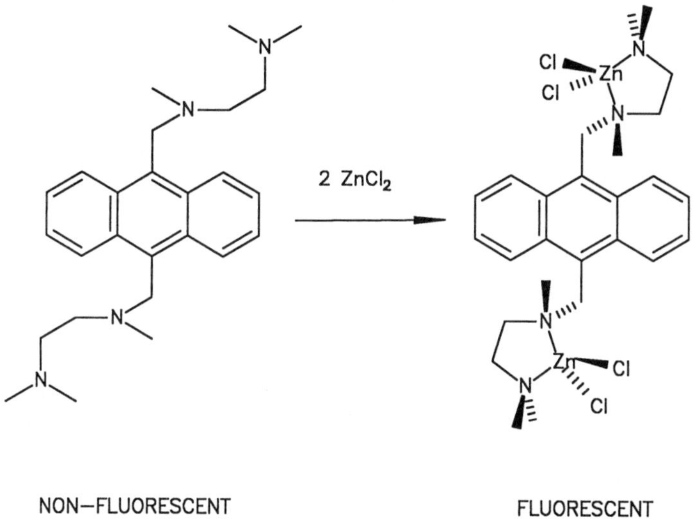

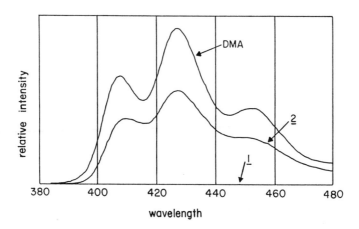

NON-FLUORESCENT

FLUORESCENT

1

2

Figure 2. Structures of 9,10-bis(TMEDA)anthracene (**1**) and its bis(ZnCl₂) complex (**2**). (Reproduced from reference 3i. Copyright 1988 American Chemical Society.)

Figure 3. Fluorescence emission spectra of compounds **1**, **2**, and 9,10-dimethylanthracene (DMA) in acetonitrile (all 10^{-4} M solutions). (Reproduced from reference 3i. Copyright 1988 American Chemical Society.)

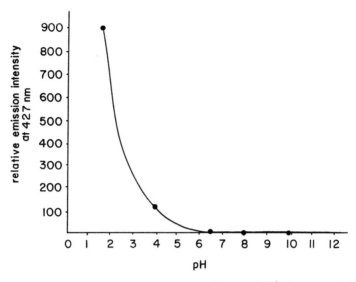

Figure 4. Fluorescence emission intensity at 427 nm of 10^{-4} M compound **1** in water as a function of pH. (Reproduced from reference 3i. Copyright 1988 American Chemical Society.)

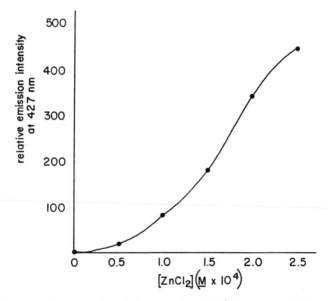

Figure 5. Fluorescence emission intensity of 10^{-4} M compound **1** in acetonitrile as a function of added $ZnCl_2$. (Reproduced from reference 3i. Copyright 1988 American Chemical Society.)

These results suggested that nitrogen-containing ligands with known specificities for metal ions might be utilized as fluorescent chemosensors via a simple, flexible connection to a fluorescent compound such as anthracene. In this scheme, complexation need not change the conformation of the fluorophore as required by an inhibition of vibrational decay mechanism, but it must tie up amine lone pairs. It also seemed possible to tie up amine lone pairs in other interesting ways, such as ion-pairing or strong hydrogen bonding; each interaction suggests potential analytical applications. We have concluded that this approach to the design of fluorescent analytical chemosensors has considerable potential, and have examined the utility of the CHEF method in other media and systems.

Chelation-Enhanced Fluorescence of Anthrylazamacrocycle Chemosensors in Aqueous Solution

While conjugate probe methods offer considerable potential and flexibility in the design of metal-selective fluoroionophores, large fluorescence changes to date have been seen only in non-aqueous solution. Of course, assays in totally aqueous media are better suited to many potential applications. Accordingly, we have examined the occurance of CHEF effects in totally aqueous environment using anthrylazamacrocycle chemosensors. Anthrylazamacrocycles **5a-e** were synthesized (Figure 6) by the reaction of 9-chloromethylanthracene (**3**) with an excess of the appropriate azamacrocycle (**4, x=1-4**). The free bases of **5a-e** were obtained as oils after selective basic extraction. All compounds were subsequently isolated as HCl salts by precipitation from ethanol with concentrated HCl. Integration of the ^{1}H NMR spectrum in each case demonstrated a 1:1 ratio of anthracene to azamacrocycle protons, confirming that overalkylation had not occurred (*6*).

The pH-fluorescence profiles for compounds **5a-e** are shown in Figure 7. We postulate that protonation at the benzylic nitrogen in all five anthrylazamacrocycles is the key step leading to large fluorescence enhancements; this idea is supported by the studies of Thomas (*7*) and of Davidson (*8*) on the distance dependance of exciplex formation in a series of homologous naphthyl- and anthrylalkylamines. While fluorescence quenching is expected at high pH, we also observed some form of quenching under strongly acidic conditions; such a decrease has been observed using other fluorophores, and in these cases is due to an acid-catalyzed, photochemically-induced decomposition. Fluorescence titrations with Zn(II) and Hg(II) (net heavy atom quenching) are shown in Figure 8. Both chelation-enhanced fluorescence (CHEF) and chelation-enhanced quenching (CHEQ) are observed for these ions, with overall emission changes of 25-fold and 20-fold, respectively. At pH 12, the presence of a large (100,000-fold) excess of added $NaClO_4$ does not interfere with the titration of Zn(II) ion. The absorption spectra remain virtually unchanged in the Zn(II) titration experiments. In an effort to test the limits of signal range in aqueous solution, we carried out the titration of **5c** with Cd(II) under strongly basic conditions (pH 13); because of the very low background fluorescence at this pH, a CHEF effect of 190-fold was obtained upon addition of a saturating amount (2 equivalents) of $Cd(ClO_4)_2$.

An observed fluorescence dependence on pH is in keeping with an intramolecular amine quenching mechanism. Protonation of an amine group in

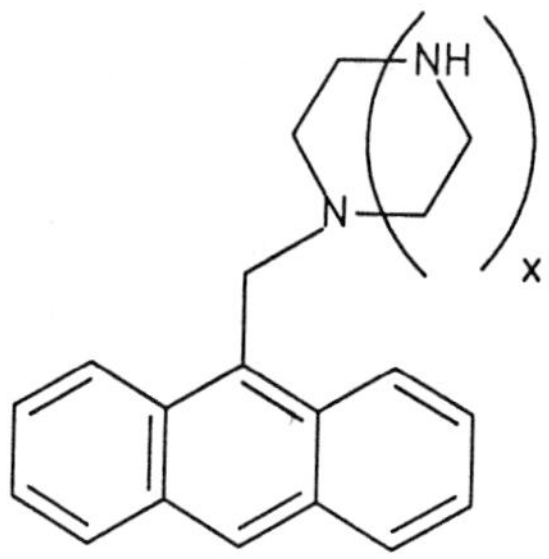

5a, x = 1
5b, x = 2
5c, x = 3
5d, x = 4
5e, x = 5

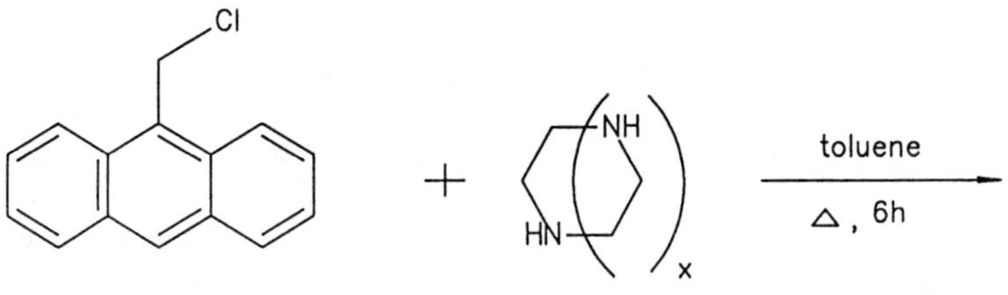

Figure 6. Syntheses of anthrylazamacrocycles **5a-e**. (Reproduced from reference 6. Copyright 1990 American Chemical Society.)

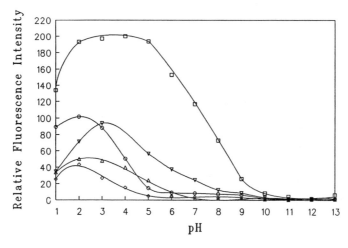

Figure 7. pH-Fluorescence profiles for 10 μM solutions of anthrylazamacrocycles **5a** (O), **5b** (▲), **5c** (□), **5d** (◊), and **5e** (▽). Excitation was at 335±3 nm; emission was measured at the emission maximum centered near 416 nm. pH's were maintained using the following solutions (all 0.1 M): trichloroacetate (pH 1); dichloroacetate (pH 2); chloroacetate (pH 3); acetate (pH 4 and 5); MES (pH 6); HEPES (pH 7 and 8); CHES (pH 9); CAPS (pH 10 and 11); tetrabutylammonium hydroxide (pH 12); NaOH (pH 13). (Reproduced from reference 6. Copyright 1990 American Chemical Society.)

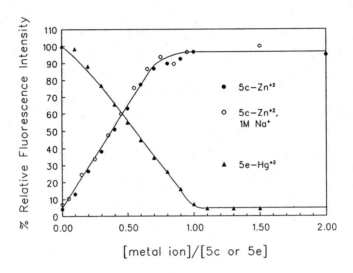

Figure 8. Titration of the fluorescence of anthrylazamacrocycles **5c** and **5e** (10 μM and 100 μM respectively) by added metal ions. The three experiments shown are: (●) titration of **5c** in pH 12 buffer with Zn(ClO₄)₂; (○) titration of **5c** in pH 12 buffer with Zn(ClO₄)₂ and 1 M NaClO₄; (▲) titration of **5e** in pH 6 buffer with Hg(ClO₄)₂. The CHEQ of **5e** by Hg(II) ion was accomplished at pH 6 because, while the azamacrocycle is sufficiently deprotonated at this pH to bind Hg(II), it is also sufficiently protonated to display some background fluorescence (Figure 7). Excitation was at 335±3 nm; emission was measured at the emission maximum centered near 416 nm. (Reproduced from reference 6. Copyright 1990 American Chemical Society.)

fluorophore-amine conjugates results in the elimination of photoinduced electron transfer. Therefore, fluorescence is expected to be a function of pH, and pH measurement using anthrylamines has been described by de Silva (*9*). The maximal emission intensity of equimolar solutions of **5a-e** at pH 2-3 varies significantly with **5c** clearly the most fluorescent; this interesting but unexplained finding is the focus of current study. In addition, it is noteworthy that compound **5c** functions as a fluorescent pH indicator with a nearly linear response between pH 5 and 10, although its metal binding properties make **5c** nonideal for this purpose.

Our survey of CHEF effects with various anthrylazamacrocycles and metal ions has permitted us to put potential interactions into three catagories:

Case 1: Complexation with fluorescence enhancement. If the metal ion itself is not quenching, binds to the azamacrocyle, and does not form a complex capable of absorbing the emitted light, a large CHEF effect (25 to 190-fold, depending on the metal, the anthrylazamacrocycle, and the pH used) is observed. Zn(II) and Cd(II) display this type of behavior. A Zn(II) titration of **5c** demonstrates a linear response as shown in Figure 8; nearly complete complexation is observed on addition of one equivalent of Zn(II) to a 10 μM solution of **5c**.

The CHEF effects we observe on the binding of Zn(II) and Cd(II) are significantly smaller than the >1000-fold enhancement we saw in acetonitrile solution (*3i*). We have observed that the background fluorescence of non-complexed anthrylpolyamines increases as the hydrogen-bonding capability of the solvent increases. For example, the background fluorescence of a 10 μM solution of anthrylazamacrocycle **5c** in pH 12 buffer is 230% that of an equimolar solution of the free base of **5c** in acetonitrile. This sensitivity to protic solvent is quite predictable; hydrogen bonding of the benzylic nitrogen will also serve to remove electron density from that position and reduce electron transfer quenching. Thus, fluorescence becomes a potentially useful tool for studying the microenvironment about the benzylic nitrogen.

Case 2: Complexation with intracomplex quenching. If a quenching metal ion (e.g., open-shell, paramagnetic, large or easily reducible cation) binds tightly to the anthrylazamacrocycle derivative, intracomplex quenching takes place (*10*). The fluorescence of compounds **5b-e** can be titrated down as shown in Figure 8 for **5e** and Hg(II). A higher concentration of the anthrylazamacrocycle is used in this case to provide sufficient emission even in the last 20% of the titration.

Case 3: No complexation. If the binding interaction is not strong enough there is no effect on the fluorescence. The fluorescence intensity of solutions of piperazinyl derivative **5a** does not change on addition of metal ions. Ca(II) and Al(III), both very weak binders to the macrocyclic polyamines (*11*), have no effect on the fluorescence of any anthrylazamacrocycle. Na$^+$ also has no effect, and as shown in Figure 8 the titration profile of **5c** with Zn(II) is virtually unaffected even by the presence of a 100,000-fold excess of Na$^+$ at pH 12.

In summary, we have shown that conjugate fluorescent chemosensors, fluoroionophores in which the metal ligand is not an integral part of the aromatic π-system, demonstrate large (20 to 190-fold) signal changes on transition metal ion binding in 100% aqueous solution. The pH dependence of fluorescence emission intensity is consistent with the elimination of photoinduced electron transfer <u>via</u>

amine protonation, and it seems likely that CHEF effects on the binding of Zn(II) and Cd(II) result from a similar mechanism. Under some conditions, the presence of 1 $\underline{M}$ sodium ion does not interfere with the titration of $\mu\underline{M}$ amounts of Zn(II). The binding of inherently quenching metals, such as Cu(II) and Hg(II), results in intracomplex quenching even though both ions are also expected to complex to the quenching amine. In that only a benzylic amine results in a nonemissive exciplex, it seems likely that it is principally the benzylic amine that is responsible for quenching of the uncomplexed anthrylazamacrocycles. Hydrogen bonding to the benzylic amine also results in increased fluorescence, although to a lesser extent than protonation or metal ion complexation.

Chelatoselective Fluorescence Pertubation in an Anthrylazamacrocycle CHEF Sensor.

While fluorimetric methods for the determination of some metal ions in aqueous solution exist using intrinsic probes (*12*), selective methods for the determination of Zn(II) and Cd(II) do not. As described in the preceeding section, anthrylazamacrocycle chemosensors **5b-e** yield large (20 to 190-fold) changes in fluorescence upon metal ion complexation in aqueous solution (*6*); the very large association constants between several transition metals (e.g., Pb(II), Cu(II), Zn(II), Cd(II), Hg(II)) and azamacrocycles (*13*) make the sequestration (and therefore quantitation) of small amounts of such ions possible. Only Zn(II) and Cd(II) bind anthrylazamacrocycles with net chelation enhanced fluorescence (CHEF); however, assigning an enhancement to one metal or the other has not been possible heretofore. We have observed that the complexation of Cd(II) and anthrylmethylpentacyclen (**5d**) uniquely demonstrates a perturbation of the fluorophore emission spectrum; the resulting ion discrimination can be utilized directly for simultaneous Zn(II)/Cd(II) analysis.

Normalized emission spectra for the complexes of Zn(II) and Cd(II) perchlorates with anthrylazamacrocycles **5b-e** (Figure 6) are shown in Figure 9. In each case an anthracenic fluorescence spectrum is observed. However, the Cd(II)-**5d** complex displays an additional broad, red-shifted band yielding the composite spectrum with λ_{max} 446 nm. A typical anthracenic emission is observed for Zn(II) and Cd(II) complexes of (9'-anthrylmethyl)-1,4,7,10,13-pentaazatridecane, a linear analog of **5d**. The unique fluorescence behavior of the Cd(II)-**5d** complex along with the large binding constants of Zn(II) and Cd(II) to pentacyclen (*13a,b*) allow for the simultaneous determination of each metal ion in aqueous solution. When the total Zn(II) and Cd(II) concentration is less than that of the probe (10 μM), the concentrations of Zn(II) and Cd(II) can be expressed as in eq. 1 and 2 .

$$[Zn(II)] = 0.54 \ I_{398} - 0.62 \ I_{516} - 1.2 \qquad (1)$$
$$[Cd(II)] = 1.1 \ I_{516} - 0.037 \ I_{398} - 0.11 \qquad (2)$$

In addition to a fluorescence pertubation, the Cd(II)-**5d** combination also uniquely yields a perturbation in the UV spectrum. A difference spectrum obtained by subtracting a fractional amount of an uncomplexed **5d** spectrum from the perturbed spectrum is the mirror image of a fluorescence difference spectrum obtained by similar means. Moreover, excitation at 400 nm (where **1-4** are weakly absorbing but where moderate absorption is seen in the difference spectrum) gives rise to an emission spectrum with identical shape and λ_{max} (456 nm) to that of the fluorescence difference spectrum. Thus, evidence points to the existence of two equilibrating ground state species as the physical basis for the chelatoselective emission. Bouas-Laurant has reported a related observation in methanol where a red-shifted CHEF was observed for a Tl(I) π-complex (*14*).

The ^{1}H NMR spectrum of the Cd(II)-**5d** complex in D_2O solution reveals the presence of more than one chelate in solution, and H-10 is clearly represented by major and minor singlets (ratio=6:4). All aromatic peaks are noticeably broadened with respect to a spectrum of **5d** taken in the absence of Cd(II) or in the presence of Zn(II). Stepwise heating of the solution (300 to 360 K) leads to a gradual coalescing of the aromatic resonances. The spectra of all other permutations of **5b-5e** with Zn(II) and Cd(II) show only one species, qualitatively similar to that of the Zn(II)-**5d** complex. Most interestingly, the Cd(II)-**5d** complex uniquely experiences deuterium exchange (k=1.5x10^{-3} sec^{-1} at 343 K and pD 8), occurring only at the 1 and 8 positions.

The observed fluorescence, UV, and ^{1}H NMR perturbations of the Cd(II)-**5d** complex are observed only in water; in methanol, ethanol, and acetonitrile only unperturbed anthracenic spectra are observed.

These observations are most consistent with the equilibria shown in Figure 10. (1) The Cd(II)-**5d** complex uniquely populates a conformer in which an anthracene-Cd(II) π-d orbital interaction is enforced. (2) The π-complex leads to a higher energy σ-complex that results in deuterium exchange. (3) Solvation strongly influences the position of the conformational equilibrium; the complete selectivity for water vs. methanol argues for stringent external ligand steric requirements in the π-complex. Arylcadmium species are well known, but previously restricted to anhydrous environments (*15*). In fact, cadmium's position above mercury in the periodic table portends activity as an electrophile towards aromatics. Conclusions regarding the structural basis of chelatoselective fluorescence perturbation suggest variations on the incorporation of such "non-classical" selectivity into future fluoroionophores.

Chelation-Enhanced Fluorescence Detection of Non-metal Ions

Both conjugate and integral fluorescent chemosensors have been applied to date almost exclusively to the detection of metal ions. Because intracomplex protonation or hydrogen bonding at a benzylic nitrogen are each expected to result in CHEF, we examined the interaction of several anions with anthracene-based conjugate fluorescent chemosensors. Anthrylpolyamines **6** and **7** (Figure 11) were synthesized by the reaction of tris(3-aminopropyl)amine with 9-chloromethylanthracene and 9,10-bis(chloromethyl)anthracene, respectively; both compounds were isolated and characterized as their HCl salts (*16*).

As described previously, examination of the literature leads to the conclusion that a change in protonation or chelation state of a benzylic nitrogen leads to large fluorescence enhancements. Thus, trication **8** (Figure 12) is the ionic form of **6**

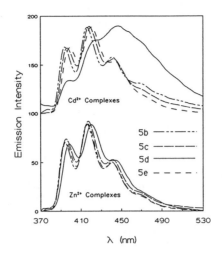

Figure 9. Normalized fluorescence spectra of 10 μM solutions of **5b-e** containing Cd(ClO$_4$)$_2$ (above) and Zn(ClO$_4$)$_2$ (below) in 0.1 M pH 10 CAPS buffer. Excitation was at 335 ± 3 nm. (Reproduced from reference 24. Copyright 1990 American Chemical Society.)

no Ar–Cd(II) complex   Ar–Cd(II) π–complex

Ar–Cd(II) σ–complex

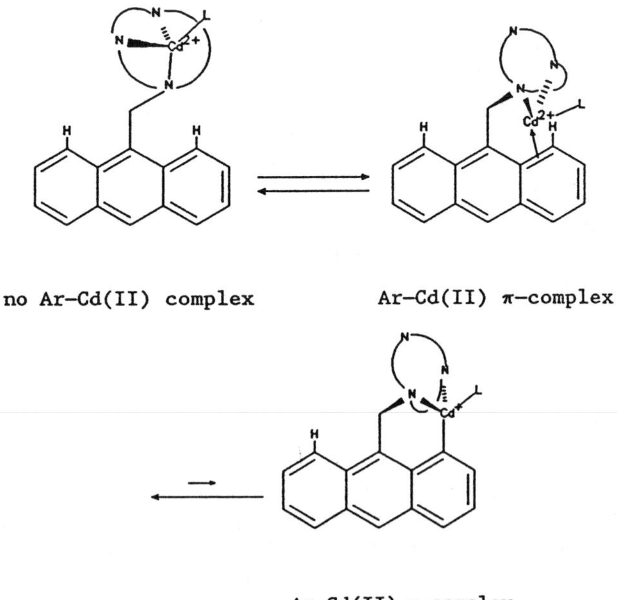

Figure 10. A proposal rationalizing the observed fluorescence pertubation and selective deuteration of the **5d**-Cd(II) complex. (Reproduced from reference 24. Copyright 1990 American Chemical Society.)

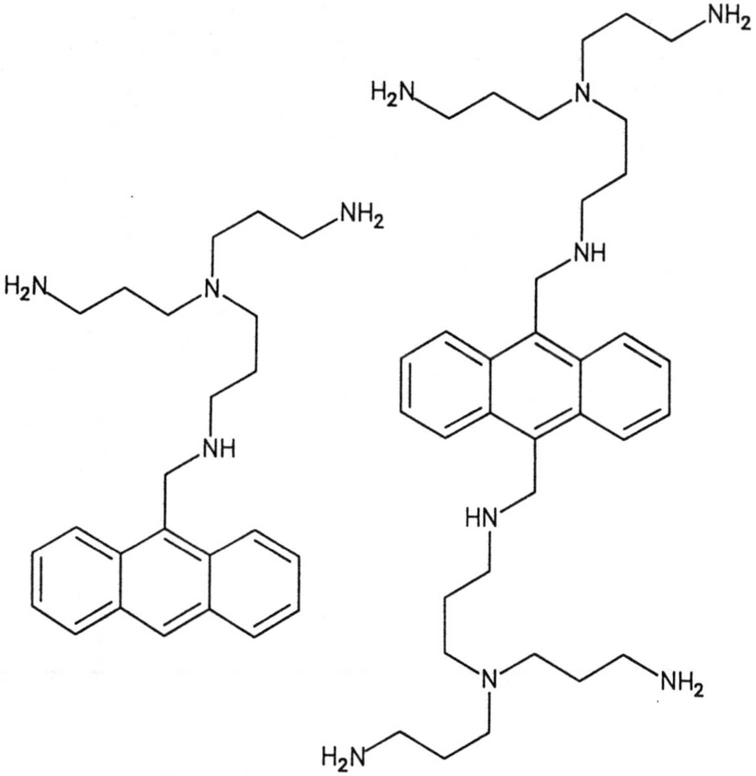

Figure 11. Structures of anthrylpolyamines **6** and **7**. (Reproduced from reference 16. Copyright 1989 American Chemical Society.)

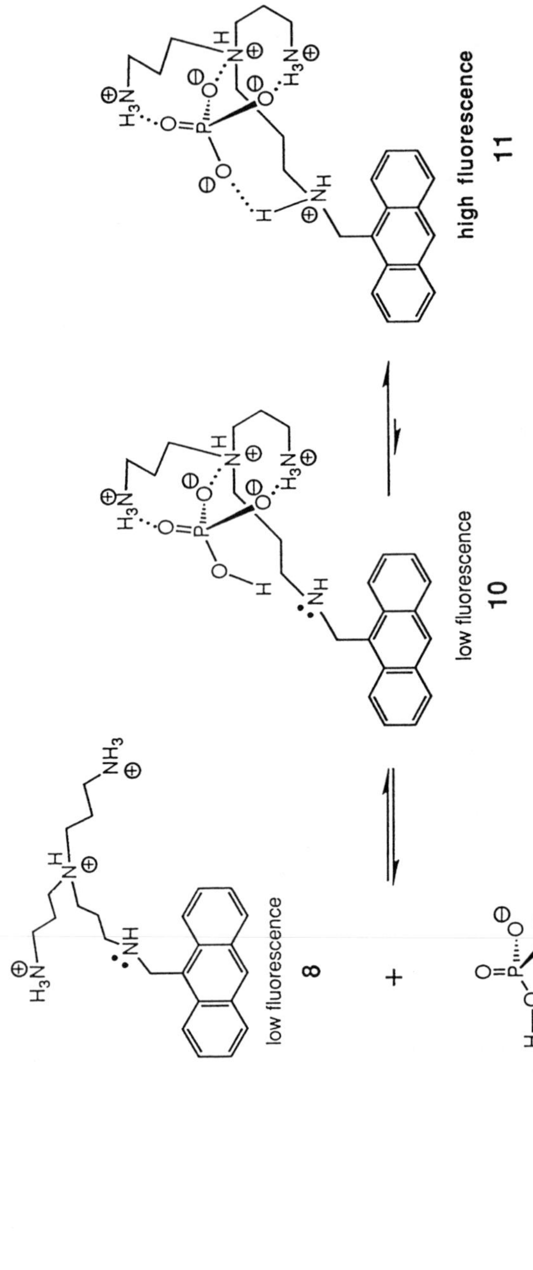

Figure 12. A working hypothesis as to how the binding of phosphate (9) to anthrylpolyamine 8 can afford a CHEF effect. Intracomplex protonation (or strong hydrogen bonding) of 10 by the HPO_4^{-2} ion leads to loss of the benzylic nitrogen lone pair (11), thereby removing this source of intramolecular quenching. (Reproduced from reference 16. Copyright 1989 American Chemical Society.)

that can act as an anion sensor at pH 6. Complexation of the anionic phosphate oxygens with ammonium ions on **8** places the remaining phosphate OH group in close proximity to the free amine; this species (**10**) will demonstrate low fluorescence due to quenching by the free amine group. However, favorable intracomplex proton transfer will lead to **11** in which intramolecular quenching is eliminated and higher fluorescence is observed. Our analysis using the HPO_4^{2-} ion as shown in Figure 12 illustrates the general principle by which CHEF is seen for anions.

Species such as sulfate and acetate, which yield smaller fluorescence enhancements upon binding, are totally dissociated at pH 6 and thus cannot deliver a proton directly; however, the observation of CHEF in these complexations does not refute the mechanism put forward in Figure 12. Intracomplex protonation (or strong hydrogen bonding) of **10** by the HPO_4^{-2} ion as shown in Figure 12 is indistinguishable from the binding of **8** to the PO_4^{-3} ion with subsequent enhanced benzylic amine protonation from the solvent. Such enhanced protonation could result from either: (a) enhanced basicity of amine groups upon ion-pairing of neighboring ammonium ions to the anion, or (b) enhanced amine protonation resulting from intracomplex hydrogen bonding of the ammonium ion to a nearby guest oxygen (e.g., **11**). If a significant concentration of PO_4^{-3} was present at pH 6, one would anticipate a CHEF effect upon its binding to **8**. Fluorescence enhancement upon the binding of acetate, sulfate, or dimethyl phosphate can likewise result from effects (a) and (b), which differ from that shown in **11** principally in the basicity of the bound anion.

Fluorescence spectra were recorded for aqueous solutions (all adjusted to pH 6.0) of **6** in the presence of increasing concentrations of several anions (Figure 13). Using the above model the concentration of complex **11** was calculated for each solution of known total anion concentration. The effective binding constant of phosphate (an average of all ionic forms) to **8** (log K_{eq}=0.82) at pH 6 along with the corresponding percent fluorescence intensity increase (>145%) could then be obtained. Similarly, values were determined at pH 6 for the binding of **6** to ATP (log K_{eq}=4.2, 79%), citrate (log K_{eq}=2.3, 97%), sulfate (log K_{eq}=1.6, 114%), acetate (log $K_{eq} \leq$0.6, >98%), and dimethyl phosphate (log $K_{eq} \leq$0.5, >66%). As an indication that even larger fluorescence enhancements are likely with structurally modified conjugate chemosensors, we have observed a 6-fold CHEF effect for the binding of citrate to anthrylbispolyamine **7** (Figure 14). These results demonstrated that intracomplex protonation of a quenching nitrogen leads to CHEF effects in aqueous solution in the same way that metal ion chelation does. We believe our results suggest a general and heretofore undescribed method for the chromogenic "signalling" of anion binding. Since the origin of the effect can be rationalized at the molecular level, a structural basis exists for the design of conjugate chemosensors for ionic and hydrogen bonding guests. Given the almost limitless synthetic approaches to nitrogen-containing hosts (*17*), the fabrication of useful analytic tools seems likely to result.

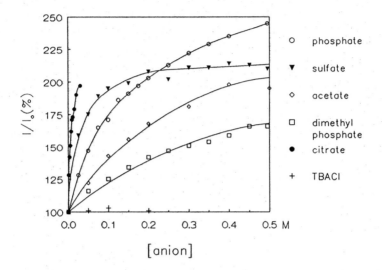

Figure 13. Titration of the fluorescence of anthrylpolyamine **6** by various added anions.

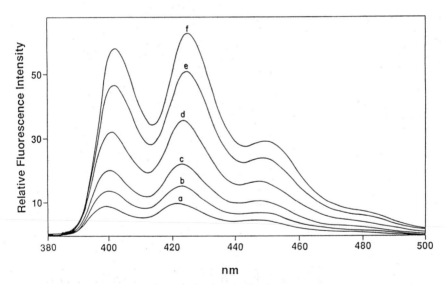

Figure 14. Fluorescence spectra for $5\mu\underline{M}$ solutions of anthrylpolyamine **7** at pH 6 in the presence of increasing concentrations of citrate: (a) none, (b) 0.1 m$\underline{M}$, (c) 1 m$\underline{M}$, (d) 10 m$\underline{M}$, (e) 0.1 $\underline{M}$, (f) 0.2 $\underline{M}$. Excitation was performed at 335 ± 3 nm. All solutions were adjusted to pH 6. (Reproduced from reference 16. Copyright 1989 American Chemical Society.)

An Assay for Enzyme-Catalyzed Polyanion Hydrolysis Based on Template-Directed Excimer Formation

While the hydrolyses of DNA and RNA can be followed by monitoring changes in their UV absorption spectra, the hydrolyses of biological polyanions lacking a chromophore must be accomplished indirectly. Both radiolabeling techniques and coupled enzyme systems are used frequently as indirect methods for enzyme assay. We have found that the activity of hydrolytic enzymes acting on the polyanions heparin and polytglutamate can be monitored by fluorescence using a template directed excimer formation effect obtained with anthrylpolyamine conjugate chemosensors.

Anthrylpolyamines **12-15** (Figure 15) were prepared via simple substitution reactions of 9-(chloromethyl)anthracene. The full emission spectra of **12-15** (all 1 μM) were collected during titration with ds DNA, ss DNA, heparin, and poly-L-glutamate; representative titration data from the monitoring of compound **14** at 422 nm are shown in Figure 16. Both disubstituted anthracenes (**14** and **15**) exhibit a 6-nm red shift in λ_{max} when bound either to ds DNA or to ss DNA; likewise, both monosubstituted anthracenes (**12** and **13**) show a 14-nm red shift in their emission spectra in the presence of either ds or ss DNA. Interaction of the nucleotide bases with the anthracene is a likely source of the bathochromic shift and the observed CHEQ effect. Such π-stacking with ss DNA, seldom observed with intercalating compounds, may result from the favorable entropy effect of electrostatic preassociation (*18*).

The heparin and poly-L-glutamate titrations show a markedly different behavior than do the DNA titrations. As polyanion is added, the fluorescence of the anthrylpolyamine solution decreases until a well-defined minimun is reached. A new emission at 510 nm, which we assign to the anthracene excimer of **14**, increases and decreases coincidently with the titrated fluorescence minimun. Likewise, the UV spectrum of 10 μM **14** with added heparin shows hypochromism that occurs and disappears coincidently with the fluorescnece minimum and a 2-nm red shift. We have proposed template-directed excimer formation as the physical basis for these observations. In the absence of heparin, fluorescence of the unassociated probe is observed. As heparin is added, the fluorescence decreases as a result of heparin-directed interaction between probe molecules. Additional heparin permits the fluorophore population to diffuse over the length of the polyanion, thus avoiding excimer formation and yielding a net CHEF.

The anthrylpolyamine most effective in binding to heparin (**14**) was used to follow the activity of heparinase at pH 5. Samples were prepared containing 1 μM probe **14** and 5 μM heparin in 0.1 M pH 5 NaOAc buffer with 0.05 mM EDTA. Under these conditions, the fluorescence of probe **3** is at its minimum as a result of template-directed excimer formation. Addition of heparinase, an enzyme that hydrolyzes heparin to oligosaccharide units (*19*), results in a flurescence enhancement as shown in Figure 17. One of the most effective polyglutamate binders (**13**) was used to test the activity of pronase at pH 5. Samples were prepared containing 1 μM probe **13** and 90 μM poly-L-glutamate in 0.1 M pH 5 NaOAc buffer with 0.05 mM EDTA. Under these conditions, the fluorescence of probe **13** is also at its minimum. Addition of pronase, a nonselective proteolytic

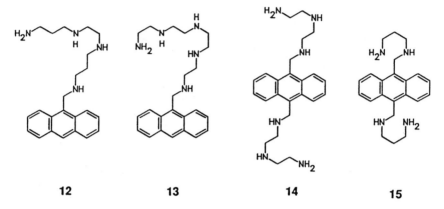

12   **13**   **14**   **15**

Figure 15. Structures of anthrylpolyamines **12-15**. (Reproduced from reference 25. Copyright 1990 American Chemical Society.)

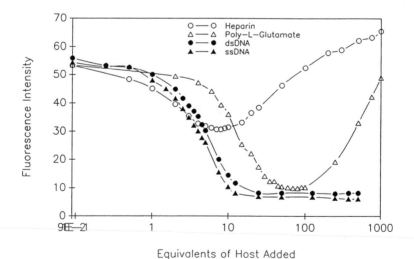

Equivalents of Host Added

Figure 16. Effects on the fluorescence of anthrylpolyamine **14** upon titration by four biological polyanions. (Reproduced from reference 25. Copyright 1990 American Chemical Society.)

enzyme that hydrolyzes polyglutamate to glutamic acid (*20*), results in a fluorescence enhancement (Figure 18). Our rationale for the physical basis of these assays is shown in Figure 19. The cationic fluoroionophores, which do not associate appreciably in dilute solution, are brought into proximity upon binding to the polyanion; excimer formation is thus enforced, and the fluorescence of the probe is quenched. Addition of hydrolytic enzyme causes cleavage of the substrate into fragments that no longer enhance probe aggregation. Consequently, quenching of the fluorescence decreases with time until hydrolysis of the template is complete. Because the binding interaction is principally electrostatic, it should be possible to follow the hydrolyses of other polyanions by this technique as well. Furthermore, it is not even necessary to know the structure of the polyanion for the assay to be useful. For this reason, we believe that the template directed excimer formation method may prove useful for the assay of other enzymes involved in the anabolism or catabolism of anionic biopolymers.

Fluorescence Chemosensing of Carbohydrates

Few chemical sensing mechanisms (other than bulk quenching) have been described for neutral molecules; nevertheless, many small analytes of interest are uncharged: glucose, for example. Of course, most carbohydrates are neither fluorescent nor are they fluorescence quenchers; novel signal transduction mechanisms are required.

   Anthrylboronic acid **16** in water displays a fluorescence emission centered at 416 nm of similar structure to that displayed by anthracene itself. Upon addition of base, fluorescence decreases; because the emission can be modulated reversibly, the change is due not to decomposition but to ionization leading to boronate **17** (Figure 20). While there are several reasons that the fluorescence of boronate **17** might be quenched compared with that of **16**, the oxidizability of borates suggests electron transfer quenching; photoinduced electron transfer from alkyltriphenylborate salts with resulting fluorescence quenching has been described by Schuster (*21*). The pH-fluorescence profile of **16** obtained without buffer is shown in Figure 21 (L), from which a pK_a of 8.8 is calculated; this compares favorably to the known phenylboronic acid pK_a of 8.83 (*22*).

   Upon addition of fructose, the apparent pK_a value decreases, leading to the remaining four curves shown in Figure 21. The explanation for this observation lies in the fact that the fructose complex of **16** is a stronger acid than is **16** itself. This result was predicated on the work of Edwards, who reported the same trend in polyol complexes of phenylboronic acid in 1959 (*23*). As determined in the presence of a near-saturating amount of fructose (100 mM), the apparent pK_a of the fructose-**16** complex is 5.9. The greatest signal range available is therefore at a pH that is the average of pK_a(**16**) and pK_a(**16·fructose**), or 7.35. Titrations of selected polyols, determined at pH 7.4, are shown in Figure 22; the apparent fructose dissociation constant at that pH is 3.7 mM. The stability trends we observe are in good agreement with the reported trends using phenylboronic acid in water (*23*). Chelation-enhanced quenching (**CHEQ**) obtained on polyol binding at constant pH results, in essence, from a shifting of the Figure 20 equilibrium from **16** (higher fluorescence) towards **19** (lower fluorescence).

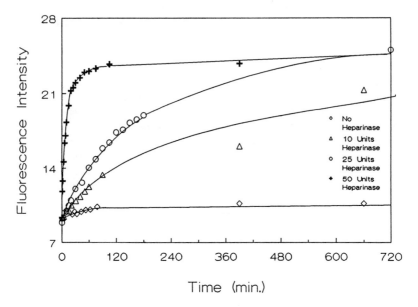

Figure 17. Effect of heparinase on the CHEQ of probe **13** with 5 equivalents of heparin. (Reproduced from reference 25. Copyright 1990 American Chemical Society.)

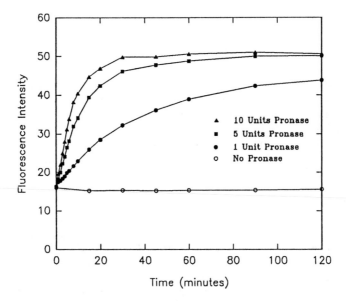

Figure 18. Effect of pronase on the CHEQ of probe **13** with 50 equivalents of poly-L-glutamate.

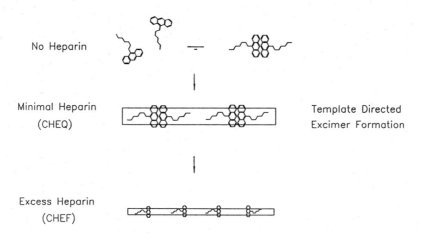

Figure 19. Proposed mechanism for template directed excimer formation. (Reproduced from reference 25. Copyright 1990 American Chemical Society.)

high fluorescence   low fluorescence

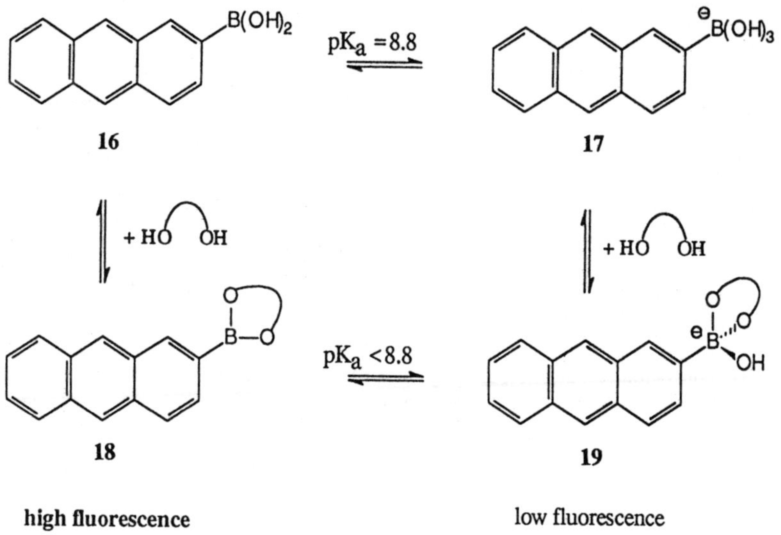

Figure 20. Equilibria available to an aqueous solution of anthrylboronic acid **16** in the presence of a generic polyol. Species **16** and **18** display higher fluorescence than do species **17** and **19**. (Reproduced from reference 26. Copyright 1992 American Chemical Society.)

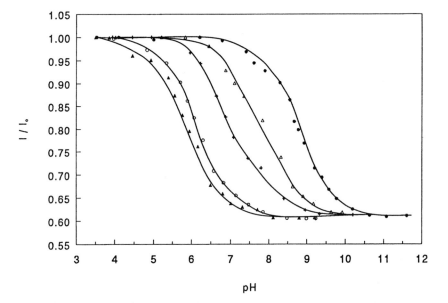

Figure 21. pH-Fluorescence titrations of anthrylboronic acid **16** (0.75 lM) as a function of fructose concentration (•, 0 mM; ▲, 2 mM; +, 5 mM; ◯, 50 mM; ▲, 100 mM). All solutions contain 1% (v/v) DMSO. The apparent pK_a's at these concentrations are, respectively: 8.8, 7.8, 7.0, 6.2, and 5.9. (Reproduced from reference 26. Copyright 1992 American Chemical Society.)

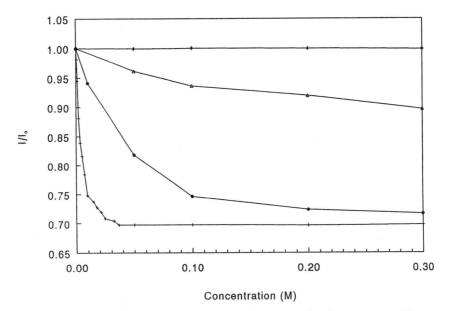

Figure 22. Fluorescence titrations of anthrylboronic acid **16** (0.75 μM) at pH 7.4 (20 mM phosphate buffer) as a function of polyol concentration (+, fructose; •, 1,1,1-tris(hydroxymethyl)ethane; ▲, glucose; +, ethylene glycol). All solutions contain 1% (v/v) DMSO. (Reproduced from reference 26. Copyright 1992 American Chemical Society.)

Conclusion

Anthrylpolyamines prove to be a synthetically accessible, yet rich source of chemosensors for a variety of ionic species in aqueous solution. The challenges that present themselves for future research in this field are intimately related to the potential applications of such compounds. Selectivity (metal ion vs. anion vs. proton) will continue to be of the greatest interest. It must be appreciated, however, that absolute selectivity is a theoretical impossibility; thus, useful selectivity ranges will be defined for particular applications. Excitation and emission wavelengths can, in principle, be engineered to avoid background absorption or autofluorescence; again, the range of usefulness must be defined prior to compound design. Down the road, the coupling of chemosensor with fiber optic methods to make remote sensing devices will create a whole new regime of questions relating to sensing on surfaces. All these issues, sitting resolutely at the interfaces of synthesis, coordination chemistry, and photochemistry, await definition and resolution.

Acknowledgments. The experimental and intellectual contributions of Drs. Xavier Cherian, Engin Akkaya, Mike Huston, Mi-Young Chae, Juyoung Yoon, Sung Yeap Hong, and Scott Van Arman are represented by the work described in this chapter.

We gratefully acknowledge support for this work from The Ohio State University, the A. P. Sloan Foundation, the Camille and Henry Dreyfus Foundation, the National Science Foundation, Merck & Co., and Eli Lilly and Company.

Literature Cited

1) (a) Schwarzenbach, G.; Flaschka, H. *Complexometric Titrations*, (translation by H. Irving), Metheun, 1969; (b) West, T. S. *Complexometry with EDTA and Related Reagents*, BDH, 1969; (c) *Indicators*, Bishop, E., ed., Pergamon, 1972; (d) Guilbault, G. G. *Practical Fluorescence*, Marcel Dekker, Inc., New York, NY, 1973, Chapter 6; (e) Clarke, R. J.; Coates, J. H.; Lincoln, S. F. Inorg. Chim. Acta **1988**, 153, 21.

2) (a) Grynkiewicz, G.; Poenie, M.; Tsien, R. Y. *J. Biol. Chem.* **1985**, *260*, 3440; (b) Tsien, R. Y. *Soc. Gen. Physiol. Ser.* **1986**, *40*, 327..

3) (a) Sousa, L. R.; Larson, J. M. *J. Am. Chem. Soc.* **1977**, *99*, 307; (b) Nishida, H.; Katayama, Y.; Katsuki, H.; Nakamura, H.; Takagi, M.; Ueno, K. *Chem. Lett.* **1982**, 1853; (c) Wolfbeis, O. S.; Offenbacher, H. *Monat. Chem.* **1984**, *115*, 647-654; (d) Konopelski, J. P.; Kotzyba-Hibert, F.; Lehn, J.-M.; Desvergne, J.-P.; Fages, F.; Castellan, A.; Bouas-Laurent, H. *J. Chem. Soc, Chem. Commun.* **1985**, 433; (e) de Silva, A. P.; de Silva, S. A. *J. Chem. Soc., Chem. Commun.* **1986**, 1709; (f) Street, K. W.; Krause, S. A. *Anyl. Lett.* **1986**, *19*, 735; (g) Ghosh, S.; Petrin, M.; Maki, A. H.; Sousa, L. R. *J. Chem. Phys.* **1987**, *87*, 4315; (h) Chemical and Engineering News, November 9, 1987, p 26; (i) Huston, M. E.; Haider, K. W.; Czarnik, A. W. *J. Am. Chem. Soc.* **1988**, *110*, 4460; (j) Fery-Forgues, S.; Le Bris, M.-T.; Guette, J.-P.; Valeur, B. *J. Phys. Chem.* **1988**, *92*, 6233; (k) Ganion, S. J.; Stevenson, R. W.; Son, B.; Nikolakaki, C.; Bock, P. L.; Sousa, L. R.; Abstract ORGN 133 from the 197th ACS National Meeting, Dallas, TX; (l) Fages, F.; Desvergne, J.-P.; Bouas-Laurent, H.; Marsau, P.; Lehn, J.-M.; Kotzyba-Hibert, F.; Albrecht-Gary, A.-M.; Al-Joubbeh, M. *J. Am. Chem. Soc.* **1989**, *111*, 8672; (m) Bourson, J.; Valeur, B. *J. Phys. Chem.* **1989**, *93*, 3871.

4) Nanjappan, P.; Czarnik, A. W. *J. Amer. Chem. Soc.* **1987**, *109*, 1826.

5) Gordon, M.; Ware, W. R., Eds. The Exciplex; Academic: New York, 1975.

6) Akkaya, E. U.; Huston, M. E.; Czarnik, A. W. *J. Am. Chem. Soc.* **1990**, *112*, 3590.

7) Chandross, E. A.; Thomas, H. T. *Chem. Phys. Lett.* **1971**, *9*, 393.

8) Brimage, D. R. G.; Davidson, R. S. *J. Chem. Soc., Chem. Comm.* **1971**, 1385.

9) de Silva, A. P.; Ripasinghe, R. A. D. D. *J. Chem. Soc., Chem. Comm.* **1985**, 1669.

10) (a) Varnes, A. W.; Dodson, R. B.; Wehry, E. L. *J. Am. Chem. Soc.* **1972**, *94*, 946; (b) Kelmo, J. A.; Shepherd, T. M. *Chem. Phys. Lett.* **1977**, *47*, 158; (c) Formosinho, S. *J. Mol. Photochem.* **1976**, *7*, 13.

11) For a review, see: Poonia, N. S., Bajaj, A. V. *Chem Rev.* **1979**, *79*, 389.

12) (a) Schwarzenbach, G.; Flaschka, H. *Complexometric Titrations*, (translated by H. Irving), Metheum, 1969; (b) West, T. S. *Complexometry with EDTA and Related Reagents*, BDH, 1969; (c) *Indicators*, Bishop, E., ed., Pergamon, 1972; (d) Guilbault, G. G. *Practical Fluorescence*, Marcel Dekkar, Inc., New York, NY, 1973, Chapter 6.

13) (a) Kodama, M; Kimura, E. *J. Chem. Soc., Dalton Trans.* **1978**, 1081; (b) Kodama, M.; Kimura, E.; Yamaguchi, S. *J. Chem. Soc., Dalton Trans.* **1980**, 2536; (c) Hancock, R. D.; Bhavan, R; Wagner, C. A.; Hosken, G. D. S. *Afr. J. Chem.* **1986**, *39*, 238.

14) Fages, F; Desvergne, J.-P.; Bouas-Laurent, H.; Marsau, P.; Lehn, J.-M.; Kotzyba-Hibert, F.; Albrecht-Gary, A.-M.; Al-Joubbeh, M. *J. Am Chem. Soc.* **1989**, *111*, 8672. An analogous ground state species was also reported for the Ag(I) π-complex; however, it is non-emissive.

15) (a) Atwood, J. L.; Berry, D. E.; Stobart, S. R.; Zorworotko, M. *J. Inorg. Chem.* **1983**, *2*, 3480; (b) Osman, A.; Steevensz, R. G.; Tuck, D. G.; Meinema, H. A.; Noltes, J. G. *Can. J. Chem.* **1984**, *62*, 1698.

16) Huston, M. E.; Akkaya, E. U.; Czarnik, A. W. *J. Am. Chem. Soc.* **1989**, *111*, 8735.

17) For an excellent overview of the great variation achievable in the design of polyammonium receptors, see: Schmidtchen, F. P. *Nachr. Chem., Tech. Lab.* **1988**, *8*, 10.

18) Previous work on polyamine DNA probes has been reported by Gabbay and by Ware: (a) Gabbay, E. J. *J. Am. Chem. Soc.* **1969**, *91*, 5136; (b) Ware, E. R.; Klein, J. W.; Zero, K. *Langmuir* **1988**, *4*, 458.

19) (a) Linhardt, R. J.; Cohen, D. M.; Rice, K. G. *Biochemistry* **1989**, *28*, 2888; (b) Linhardt, R. J.; Fitzgerald, G. L.; Cooney, C. L.; Langer, R. *Biochim. Biophys. Acta* **1982**, *702*, 197.

20) Smyth, D. G. *Methods Enzymol.* **1967**, *11*, 214.

21) (a) Chatterjee, S.; Davis, P. D.; Gottschalk, P.; Kurz, M. E.; Sauerwein, B.; Yang, X.; Schuster, G. B. *J. Am. Chem. Soc.* **1990**, *112*, 6329; (b) Schuster, G. B. Pure Appl. Chem. **1990**, 62, 1565. One referee has noted that, because the quenching we observe is less than that in these references, mechanistic statements in the present instance should await further study.

22) (a) Nakatani, H.; Morita, T.; Hiromi, K. *Biochim. Biophys. Acta* **1978**, *525*, 423; (b) Juillard, J.; Geugue, N. *C. R. Acad. Paris C* **1967**, *264*, 259.

23) Lorand, J. P.; Edwards, J. D. *J. Org. Chem.* **1959**, *24*, 769.

24) Huston, M. E.; Czarnik, A. W. *J. Am. Chem. Soc.* **1990**, *112*, 7054.

25) Van Arman, S. A.; Czarnik, A. W. *J. Am. Chem. Soc.* **1990**, *112*, 5376.

26) Yoon, J.-Y.; Czarnik, A. W. *J. Am. Chem. Soc.* **1992**, *114*, 5874.

RECEIVED June 14, 1993

Chapter 9

Fluorescent and Photochemical Probes of Dynamic Biochemical Signals inside Living Cells

Roger Y. Tsien

Howard Hughes Medical Institute and Departments of Pharmacology and Chemistry, University of California—San Diego, La Jolla, CA 92093-0647

A particularly popular and powerful use of fluorescent indicators is nondestructive monitoring of the activity or free concentration of important ions or messengers such as Ca^{2+}, H^+, Na^+, Mg^{2+}, Cl^-, or cyclic adenosine 3',5'-monophosphate (cAMP) inside living cells and tissues. Successful indicators and chelators must fulfill many biological criteria in addition to the chemically obvious ones of sensitivity and specificity. Four examples will be discussed here. A family of Ca^{2+} indicators is derived from the versatile Ca^{2+}-chelator 1,2-bis(o-aminophenoxy)ethane-N,N,N',N'-tetraacetic acid by extension into rigidized stilbenes or linkage to xanthene fluorophores. To show whether complex Ca^{2+} signals are sufficient or necessary for biological functions, photolabile chelators are valuable because flash photolysis can suddenly and cleanly release or sequester the analyte. The only fluorescent indicator for Na^+ so far demonstrated to work intracellularly is "SBFI", a diaza-15-crown-5 with additional liganding methoxyls attached to fluorophores on the nitrogens. An indicator for cAMP is prepared by labeling the regulatory and catalytic subunits of cAMP-dependent protein kinase such that binding of cAMP and dissociation of the subunits is sensed by loss of fluorescence energy transfer. This protein-based indicator shows that fluorescence chemosensors can measure analytes larger and more complex than simple inorganic ions and that techniques of molecular biology and protein engineering may offer exciting new possibilities in this field.

One of the most important practical applications of fluorescent chemosensors has been in cell biology, where such indicator molecules have become the basis for the most important methods for measuring free concentrations or activities of ions and messenger substances inside living cells (1,2). Fluctuations in these concentrations are often fast and highly localized and are of tremendous

0097–6156/93/0538–0130$06.00/0

importance in biological signal transduction (3,4). Properly designed fluorescent chemosensors are among the best techniques for sensing such concentrations, for the following reasons:

1) Chemosensors that use well-defined thermodynamic equilibria for measuring analytes in living cells inherently measure the free concentrations or activities (5), which are usually much more biologically relevant and interpretable than measurements of the total concentration (bound plus free) of the analyte in dead tissue. The latter result from a wide variety of traditional destructive analytical techniques such as flame photometry, atomic absorption spectrophotometry, neutron activation analysis, radioisotope equilibration, immunoassay of cell lysates, electron-beam microprobe analysis, etc. For most tissues and many intracellular messengers, the free concentration or activity that controls other relevant biochemical equilibria is a tiny and variable fraction of the total concentration, because the total is dominated by material bound or sequestered on sites irrelevant to the particular equilibrium of interest. Therefore destructive methods for measuring total concentrations are usually unable to give reliable information about the more relevant and dynamic free concentrations, just as a combustion microanalysis for total hydrogen in a sample is a poor guide to its pH.

2) Optical techniques in general are relatively non-invasive, sensitive, fast, and easily adapted to high spatial resolution via microscopic imaging.

3) Fluorescence is one of the best modes of optical readout. It is well known that fluorescence can be enormously more sensitive than absorbance when sample pathlength and indicator concentration are very small, because fluorescence is detected against a background of nearly zero whereas small absorbances have to be detected as decrements from the high intensity of the transmitted beam. Fluorescence likewise has an inherent advantage over chemiluminescence in that each individual fluorophore can emit tens to hundreds of thousands of photons before being bleached by parasitic reactions, whereas chemiluminescent molecules emit at most one photon each before exhaustion (6). Also, fluorescence is uniquely suited to special techniques such as resonance energy transfer, which detects proximity of appropriate fluorophores at the few-nanometer range (7-10), anisotropy measurements, which detect rotational mobility at the nanosecond time scale (11), and confocal microscopy, which enables a given plane of focus to be imaged deep inside a thick specimen without physically cutting away the material in front and in back of the desired plane (1,12).

Biological Requirements for Intracellular Fluorescent Chemosensors

However, in order to make use of these advantages, the fluorescent chemosensor must satisfy a number of requirements (13), some of which may not be obvious to workers who are not cell biologists:

1) The chemosensor must bind the analyte with sufficiently rapid kinetics and a dissociation constant roughly comparable to the range of free concentrations to be measured. Obviously the affinities must be measured in aqueous solutions of appropriate ionic strength. Chemists often add organic cosolvents to increase the solubility of the chemosensor and decrease solvation of analytes, but such tricks are forbidden to cell biologists. Typical concentrations of interesting ions

and messengers in the cytoplasm of animal are listed in Table I. Excessive affinity is just as undesirable as insufficient affinity, since the chemosensor would start out already saturated with the analyte and insensitive to further changes. The need to tune the binding affinity to a specific range is more stringent for these nonperturbing measurements of free concentration than for probes intended for destructive measurements of total analyte concentration, for which the higher the affinity the better.

2) The chemosensor should discriminate adequately between the desired species and potential competitors even at the highest concentrations physiologically attained by the latter. Discrimination should ideally result from non-binding of the competitors rather than spectral differences between the various complexes, since even if complexes with competitors are spectroscopically invisible, they still depress the effective affinity for the desired analyte.

3) The fluorescence should be as intense as possible, characterizable by a product of extinction coefficient and fluorescence quantum yield exceeding 10^3 - 10^4 $M^{-1}cm^{-1}$. Photostability in the presence of dissolved oxygen is also highly desirable.

4) Excitation wavelengths should exceed 340 nm, because shorter wavelengths require expensive quartz rather than glass microscope optics and are strongly absorbed by nucleic acids and aromatic amino acids.

5) Emission wavelengths should exceed 500 nm to reduce overlap with tissue autofluorescence, which peaks near 460 nm due to reduced pyridine nucleotides.

6) Binding of the analyte should cause a large wavelength shift in the excitation or emission spectrum or both, so that ratioing of signals at two excitation or two emission wavelengths can be performed (14,15). Such ratioing is the instrumental analog of color vision and is extremely valuable in canceling out such irrelevancies as cell thickness, dye concentration, and wavelength-independent variations in illumination intensity and detection efficiency. If the chemosensor does not shift wavelengths but only gets brighter or dimmer upon binding the analyte, then changes in cell shape, indicator content, lamp brightness or detection efficiency can masquerade as changes in analyte concentration, whereas such artifacts are largely excluded by ratioing. In principle measurements of excited-state lifetime can give similar cancellation (16), but the experimental equipment is much more elaborate than for wavelength discrimination. Also, to resolve the bound and free forms of the chemosensor requires a large difference in their lifetimes, but such a large difference usually implies that the shorter-lived species will be relatively dim in fluorescence (17).

7) Binding to cellular constituents and membranes should be minimized or carefully controlled for specificity. Usually, enough charged polar groups such as carboxylates should be added to the chemosensor to render it highly water-soluble and impermeant through membranes, so that once introduced into cells it does not rapidly leak out again. Large domains of unrelieved hydrophobicity should be avoided unless the intention is to stick nonspecifically to membranes.

8) The polar groups just mentioned should be maskable by nonpolar protecting groups removable by cytoplasmic enzymes, so that large populations of cells can be loaded with the indicator by incubating them with the nonpolar,

Table I. Typical Concentrations of Interesting Ions and Messengers in the Cytoplasm of Animal

Analyte	Typical Conc. in Resting Cells[a]	Intracellular Physiological Range[b]	Representative Fluorescent Chemosensor(s)[c]	Effective Dissociation Constant[d]	Typical Excitation Wavelengths[e] (nm)	Typical Emission Wavelengths[e] (nm)	Best Detection Mode[f]	1st Biological Reference	No. of Citations 1982-1992
H^+	100 nM	10-1000 nM	DHPN BCECF SNARF-1	10 nM 107 nM 32 nM	360/420 440/490 517/576	455/512 530 587/640	Em. ratio Exc. ratio Em. ratio	(63) (64) (65)	6 290 21
Na^+	4-10 mM	0-100 mM	SBFI	18 mM	340/385	530	Exc. ratio	(13)	21
K^+	100-140 mM	20-160 mM	PBFI	100mM	340/350	530	Exc. ratio or Intensity	(13)	3
Cl^-	5-100 mM	0-100 mM	SPQ TMAPQ	83 mM 50 mM	350 355	442 450	Intensity Intensity	(66) (67)	22 1
Mg^{2+}	0.5-2 mM	0.2-5 mM	FURAPTRA= Mag-fura-2	1.5 mM	335/370	510	Exc. ratio	(68)	23
Ca^{2+}	50-200 nM	10 nM-10 µM	quin-2 fura-2 indo-1 fluo-3	115 nM 224 nM 250 nM 400 nM	340 340/380 331/349 500	505 505 405/485 530	Intensity Exc. ratio Em. ratio Intensity	(19) (26) (26) (28)	1,052 1,976 463 75
cAMP	<50 nM	0-10 µM	FlCRhR	100 nM	490	520/590	Em. ratio	(55)	

Continued on next page

Footnotes to Table I

[a] Crude estimates for the cytosol of typical mammalian cells not being stimulated or knowingly perturbed.

[b] Approximate range encountered in stimulated or reasonably perturbed cells.

[c] For structures see Figure 1. Emphasis has been placed on indicators shown to work intracellularly and whose structures and properties have been published.

[d] Effective dissociation constant for analyte from the chemosensor in mammalian cytosol or ionic media mimicking cytosol.

[e] Typical excitation and emission wavelengths used for measuring analyte. These do not necessarily correspond to spectral maxima, because observing wavelengths are often displaced to improve discrimination between spectra of complexed and free forms or separation between excitation and emission wavelengths. When a pair of wavelengths is given, the first emphasizes the fluorescence of the complexed form while the second is preferential for the free chemosensor. Excitation and emission bandwidths are often considerably widened in order to maximize detection sensitivity.

[f] The most sophisticated mode of fluorescence readout in reasonably common use. Intensity = measurement of fluorescence intensity at fixed wavelengths; Exc. ratio = measurement of fluorescence at two separate excitation wavelengths and a common emission band and ratioing of the two excitation amplitudes; Emission ratio = measurement of fluorescence at a common excitation wavelength and two separate emission bands and ratioing of the two emission amplitudes. Some intensity measurements may be replaced in the future by excited-state lifetime measurements.

[g] First reference discussing or demonstrating use in biology.

[h] Number of references found in MEDLINE (Jan. 1982 - Sept. 1992), that list the chemosensor as a keyword. Some attempt has been made to pool synonyms while excluding duplicate entries.

membrane-permeant derivative. Such use of the cytoplasm as the final deprotecting step avoids the requirement for microinjecting individual cells or disrupting their outer membranes as in liposome fusion, electroporation, or other lytic methods. The most general solution (18) is to include carboxylates protected as acetoxymethyl (AM) esters, i.e. geminal diesters between acetic acid and the acid group on the chemosensor. AM esters are believed to hydrolyze via cleavage of the acetyl ester by ubiquitous esterases. The resulting hydroxymethyl esters spontaneously eliminate formaldehyde and release the desired carboxylate. Therefore the latter does not need to be recognized by the esterase. Surprisingly, the formaldehyde causes negligible acute toxicity due to its slow generation in low concentrations and the availability of additional antidotes if necessary (1). Other labile carboxylic derivatives such as anhydrides or high-energy esters have the potential drawback that acylation of cellular nucleophiles might compete with hydrolysis, which would be bad for both the chemosensor and the cell.

9) Toxicity, both intrinsic and photodynamic, obviously needs to be minimized, at least over the time scale of the measurements, typically minutes to hours.

Satisfying all these requirements simultaneously is difficult; partial successes can still be useful. Currently the major analytes for which fluorescent chemosensors have been demonstrated to work inside living cells are H^+, Na^+, K^+, Cl^-, Mg^{2+}, Ca^{2+}, and cyclic adenosine-3',5'-monophosphate (cAMP, Fig. 1), as listed in Table I. Successful engineering of such chemosensors opens up large areas of biological exploration, as shown by the approximate numbers of citations for each chemosensor.

The remainder of this paper will focus on chemosensors for Ca^{2+}, Na^+, and cAMP.

Fluorescent Indicators for Calcium

Ca^{2+} represents the first analyte for which a binding site was specifically designed and built into chemosensors aimed at intracellular applications (19). Measurement of Ca^{2+} is still by far the most popular application of fluorescent chemo-sensors, due to the tremendous and ubiquitous biological importance of intracellular Ca^{2+} signaling (20-23). Local pulses of elevated Ca^{2+} control the force-generating machinery in every muscle in the body and the release of neurotransmitters at every chemical synapse in the nervous system. Most though not all glands in the endocrine and exocrine systems, as well as most of the cells of the immune system are strongly influenced by intracellular Ca^{2+}. Even at conception, Ca^{2+} pulses in the spermatozoon and ovum are key events in signaling their mutual recognition, promoting fusion of the two cells, and triggering subsequent development of the embryo. Many other examples could be given. It is important to remember that extracellular Ca^{2+} concentrations (1 - 10 mM) are typically four to five orders of magnitude greater than basal intracellular levels (~100 nM), so that any damage to the outer membrane of the cell causes the Ca^{2+} to rise drastically. Therefore it is particularly important to be able to get the fluorescent chemosensor into the cells without having to puncture or otherwise damage their plasma membranes.

The major chemical problem in recognizing intracellular Ca^{2+} is how to discriminate against Mg^{2+}, whose intracellular free concentration exceeds that of Ca^{2+} by about four orders of magnitude under basal conditions. A solution (19) was to build a binding site isosteric with ethyleneglycolbis(β-aminoethylether)-N,N,N',N'-tetraacetic acid (EGTA, Fig. 1), the best-known artificial chelator with sufficient ($>10^5$) Ca^{2+}-to-Mg^{2+} selectivity. Replacement of the two-CH_2CH_2-groups between the nitrogens and ether oxygens by ortho-aromatic linkages gave 1,2-bis(o-aminophenoxy)ethane-N,N,N',N'-tetraacetic acid (BAPTA, Fig. 1), which preserves the Ca^{2+}-to-Mg^{2+} specificity, apparently because the binding site retains the dimensions and large donor number appropriate for the larger Ca^{2+} ion, whereas Mg^{2+} is too small and can be coordinated by only half of the binding site at any one time (19). The aromatic nuclei confer ultraviolet absorbance and fluorescence spectra, which are strongly modulated by Ca^{2+} binding. The spectral sensitivity arises because the nitrogen lone pair of electrons conjugates with the aromatic π-orbitals in the absence of Ca^{2+} but is decoupled from the π-system when Ca^{2+} binds. This decoupling is analogous to the well known effect of protonation on anilines and is probably due both to electrostatic attraction towards the cation and to steric twisting of the nitrogen-to-aryl carbon bond. Such twisting is necessary to fold the ligand around the Ca^{2+} and has been verified by X-ray structural determinations on Ca^{2+} complexes of this family (24). Additional benefits of incorporating aromatic rings next to the nitrogens include reduction in H^+ interference due to the lower pK_a of anilines compared to aliphatic amines; acceleration of the Ca^{2+} association rate because H^+ no longer has to be ejected from the binding site at physiological pH; and tunability of the cation affinities by substituents on the aromatic ring acting through typical Hammett-type relationships (19,25).

The problem of how to couple strongly fluorescent chromophores to BAPTA has been solved in two general ways. Direct extension of the original aniline chromophore to make heterocyclic stilbene analogs (26) gives Ca^{2+} indicators such as fura-2 and indo-1 (Fig. 1) that retain good affinity for Ca^{2+}, give large shifts in near-UV excitation and/or visible emission spectra upon binding Ca^{2+}, and load quite well into a wide variety of tissues by means of their membrane-permeant acetoxymethyl esters. The limitation of this approach has been the difficulty in pushing the excitation and emission spectra to longer wavelengths. Simply extending the chromophores does not give major bathochromic shifts but tends to add excessive molecular weight and hydrophobicity, which hinder tissue permeation and loading. Increasing the electron-acceptor strength of the groups at the far end of the chromophore does give large red shifts but tends to decrease the Ca^{2+} affinity to an undesirable degree and also detracts from the fluorescence quantum efficiency in aqueous solution (5). The latter effect may be due to the increasing "push-pull" or charge-transfer character of the excited state, whose large dipole moment strongly interacts with the water dipoles to promote radiationless quenching. Unfortunately, rational design of ratiometric fluorescent indicators for Ca^{2+} has not been easy. Production and design of molecules that change their absorbance spectrum upon Ca^{2+} binding is relatively simple, but maintaining high fluorescence quantum efficiency in both the free and Ca^{2+}-bound species is a more difficult and empirical challenge. Emission ratioing

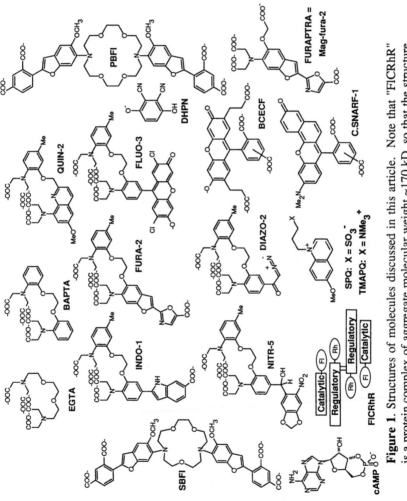

Figure 1. Structures of molecules discussed in this article. Note that "FICRhR" is a protein complex of aggregate molecular weight ~170 kD, so that the structure shown for it is highly schematic and at a quite different scale from all the other structures.

capability is yet rarer than excitation ratioing because a change in emission wavelengths requires that the Ca^{2+} remain bound in the excited state. However, most anilines have a much lower pK_a and electron density on nitrogen in their first excited singlet state than in their ground state (27), so there is a strong tendency for the Ca^{2+} to shift away from the excited state nitrogen and thus to lose its effect on emission wavelength.

An alternative approach exemplified by fluo-3 (Fig. 1) (28) (see also articles by de Silva and by Kuhn in this volume) is to link the Ca^{2+}-binding site to various standard, high-quantum-efficiency visible-wavelength fluorophores via short but mostly insulating spacers. Now the action of the electron-rich aniline groups in the absence of Ca^{2+} is to quench the excited state of the appended fluorophore by the process of photoinduced electron transfer (PET). Ca^{2+} binding discourages formation of the anilinium radical cation and inhibits quenching, i.e. raises the fluorescence quantum yield, without any significant shift in wavelengths. This approach is very flexible, in that a variety of fluorophores and spacers can be relatively quickly attached to the standard Ca^{2+} binding site. Ground-state Ca^{2+} affinity is little affected by the pendant fluoro-phore because of the insulating linkage. However, the lack of any excitation or emission shift upon Ca^{2+} binding prevents ratioing of the signals obtained at two differentially sensitive wavelengths. This is a major biological limitation that greatly hinders quantitative calibration of the fluorescence signals in terms of absolute Ca^{2+} levels (29). Also, the large size of most of the chelator-spacer-fluorophore conjugates makes intracellular loading via permeant esters more difficult, perhaps because aqueous solubility of the hydrophobic esters becomes inadequate. Because of these problems, the UV-excited, wavelength-ratioable indicators fura-2 and indo-1 are still the most widely used (Table I). Together with appropriate computerized instrumentation for acquiring images at two excitation or two emission wavelengths and ratioing their intensities, they provide dramatic images of Ca^{2+} pulses, steps, waves, and oscillations in a wide variety of cells (14,30-32).

Photolabile Chelators for Manipulating Ca^{2+} Levels

The complex spatiotemporal patterns of Ca^{2+} signals are colorful and fascinating, but to determine their biological function we need means of artificially mimicking or suppressing them to see what effects on cell physiology result. A powerful approach for raising and lowering Ca^{2+} with high spatial and temporal resolution is to build and use photolabile chelators that use light not to signal Ca^{2+} but to control it. In the strategy from our laboratory, the standard Ca^{2+}-binding site is coupled not to a fluorophore but to photoisomerizable groups *para* to the anilino nitrogen that irreversibly change their electron affinity upon illumination. For example, in "nitr-5" (Fig. 1) the nitrobenzyl substituent isomerizes and extrudes a mole of H_2O to form a nitrosobenzoyl group with much greater electronegativity, causing a 40-fold drop in Ca^{2+} affinity and extrusion of Ca^{2+} under physiological conditions (33). Conversely, in "diazo-2" (Fig. 1) the diazoacetyl substituent upon illumination extrudes N_2 to form a carbene, which undergoes a Wolff rearrangement to a ketene, which hydrolyzes to a carboxymethyl group, which is much less electron-withdrawing and allows a 30-fold increase in affinity over the starting material (34). These and other photolabile molecules from other laboratories (35,36) have been of considerable value in analyzing the physiological importance and mechanisms of various fast Ca^{2+} signals (37-39) and are of chemical relevance to the theme of this book because they use the same principles of affinity, selectivity, and quantum yield optimization to achieve a complementary goal.

Fluorescent Indicators for Sodium

Intracellular Na^+ is important in animal cells because the gradient of $[Na^+]$ between extracellular (120 to 450 mM) and intracellular (4 to 20 mM) concentrations is used to power nutrient uptake, epithelial transport, regulation of other intracellular ions, and transmission of electrical impulses. So far Na^+ does not seem to be used as a signal messenger the way Ca^{2+} is so commonly used, perhaps because the basal Na^+ concentration is high enough so that a large flux of Na^+ is required to make much of a percentage change. However, measuring the $[Na^+]$ gradient is important for understanding cellular energy metabolism and how it adjusts to changing loads. Extracellular $[Na^+]$ is usually fairly stable and easily measurable by flame photometry or ion-selective electrodes, so that intracellular $[Na^+]$ is the main technological challenge at the research biological level. So far the only optical indicator for Na^+ that has been demonstrated to work in cells is "SBFI" (13). SBFI is a crown ether with two pendant ligand arms somewhat similar to the "bibracchial lariat ethers" developed by Gokel and collaborators (40,41). However, SBFI has rigid aromatic rather than flexible aliphatic arms, considerably higher Na^+:K^+ selectivity, and fluorescence ratio responsivity to Na^+ binding, probably due to mechanisms analogous to those for Ca^{2+} binding to fura-2. Four additional carboxylates were added to confer water solubility and the possibility of loading into intact cells by hydrolysis of its membrane-permeant acetoxymethyl ester. Biological considerations were essential in choosing tradeoffs between the various properties desired for the molecule. For example, higher affinities and Na^+:K^+ selectivities were obtained by replacing the methoxyaryl groups by bulkier and stronger ligands such as quinaldines or acridines, but H^+ interference developed and fluorescence efficiency sank, so that the best balance of properties was obtained with SBFI. Merely moving the terminal carboxylates into *ortho* instead of *meta* relationships caused excessive compartmentation and labeling of intracellular proteins or organelles, perhaps due to anhydride formation (13,42). In the presence of typical vertebrate intracellular K^+ levels, the effective dissociation constant of SBFI for Na^+ *in vitro* is around 18 mM (13), which is reasonably well suited to monitor $[Na^+]_i$ changes from the typical resting level upwards. Calibration requires some care because the dye can be perturbed by cytoplasm and is still subject to some compartmentalization into organelles, and is most conveniently performed in intact cells with the pore-forming antibiotic gramicidin, which rapidly clamps $[Na^+]_i$ and $[K^+]_i$ equal to the extracellular levels of those ions (42). SBFI has revealed in cell types such as single fibroblasts (42), gastric cells (43), platelets (44), and cardiac myocytes (45) that cell activation causes much larger increases in cytosolic Na^+ (typically 50-100%) than previously suspected. Recent studies in cerebellar neurons (46,47) have shown that the Na^+

influx due to single action potentials can be detected in the central cell body and proximal dendrites, but that the Na^+ spike seems not to spread out to the more remote parts of the cell whereas the Ca^{2+} component does. Even more quantitative measurements of Na^+ influx and efflux rates in single cells (43) can be made by comparing SBFI responses in cells in normal media with those in the absence of external Na^+ or K^+ or in the presence of sodium pump inhibitors.

Fluorescent Indicators for cAMP

Until recently, simple spherical inorganic ions were the only species for which fluorescent chemosensors existed, so a legitimate concern might have been that this sensor technology might be limited to such species. We were strongly motivated to disprove this restriction. As cell biologists focusing on signal transduction, we felt the most important target was cyclic adenosine 3',5'-monophosphate (cAMP, Fig. 1), since it is the longest studied intracellular second messenger and the only one known to have a ubiquity and importance comparable to Ca^{2+} (48,49). The standard methods for quantifying cAMP all involved killing the cells, usually in very large numbers, and analyzing their contents by immunoassay. Several suggestions had been made that cAMP signals may be compartmentalized (50-53), but such localization was hard to test due to the crudity of the assays. A fluorescent indicator of cAMP that could be imaged in single living cells would therefore be of considerable interest. However, the size and shape of cAMP seemed too complex for us to design a totally artificial binding site with sufficient affinity and selectivity. Very recently an abiotic cAMP ligand has been described (54), but so far it has been characterized only by NMR in deuterochloroform, has too low an affinity by several orders of magnitude, and is of uncertain selectivity, but nevertheless represents astonishing progress and a promising start. Meanwhile, we chose instead to modify a natural cAMP-sensing protein. Which protein should be chosen and how could its fluorescence be made sensitive to the binding of cAMP? After several years of fitful ruminations and trials, an attractive and retrospectively obvious scheme (Fig. 2) was finally conceived and made to work in a collaboration involving Stephen Adams in my laboratory and Prof. Susan Taylor of the Dept. of Chemistry at UCSD. This indicator relies on the known biochemical fact that the most important or only molecule with which mammalian cells naturally sense cyclic AMP is cAMP-dependent protein kinase (cAMPdPK), which consists of a complex between regulatory (R) and catalytic (C) subunits (49). In the absence of cAMP, this complex has a stoichiometry of R_2C_2 and is inactive as a kinase, since the R subunits inhibit the kinase domains of the C subunits. Upon binding cAMP, the R subunits dissociate from the C subunits, releasing the biological activity of the latter. cAMPdPK is made into a fluorescent sensor by covalently labeling the R subunits with a fluorescent tag and the C subunits with a different fluorescent tag, such that in the intact R_2C_2 complex, light energy absorbed by one tag is transferred to the other tag by the process of radiationless energy transfer (7-10). This phenomenon requires that the wavelengths of fluorescence emission from one dye, the donor (D), should overlap the wavelengths of absorbance of the other dye, the acceptor (A), and that the D and A moieties should be in close molecular

Fluorescence Detection of Cyclic AMP

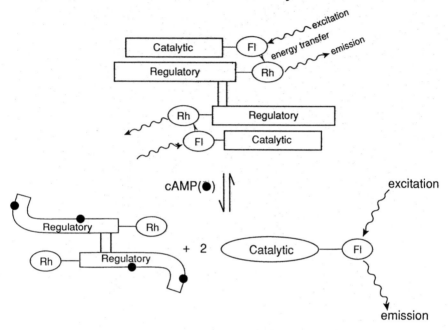

Figure 2. Schematic depiction of fluorescent resonance energy transfer (FRET) between the subunits of cAMP-dependent protein kinase (55) used for detecting cAMP. In the absence of cAMP, most of the kinase is in the holoenzyme state, which facilitates FRET between the fluorescein-labelled catalytic (C) subunits and the rhodamine-labelled regulatory (R) subunits. Excitation of fluorescein gives substantial re-emission from rhodamine at the expense of fluorescein. In the presence of high concentrations of cAMP, the R subunits undergo conformational change that dissociates them from the C subunits and prevents FRET. Therefore, excitation of fluorescein results in fluorescein emission rather than rhodamine re-emission (Reproduced with permission from ref. 55. Copyright 1991 Macmillan Magazines).

proximity to each other, < 4 - 6 nm apart. Provided suitable choices are made for D and A, this condition of close proximity is reasonably well satisfied in the holoenzyme complex, so that energy transfer occurs with considerable efficiency. Such transfer is easily detectable by excitation at wavelengths absorbed by D, since emission from D is quenched and the energy appears instead at the longer wavelengths characteristic of A's emission. However, once cAMP has dissociated R from C, D and A thereby become widely separated from each other and unable to transfer energy from D to A. Now illumination at wavelengths absorbed by D simply causes fluorescence emission from D, with no involvement of A, so that the composite emission spectrum shifts toward the shorter wavelengths of D. Currently preferred choices of dyes are D = fluorescein and A = tetramethylrhodamine, attached via isothiocyanates and N-hydroxysuccinimide labeling reagents to the catalytic and regulatory subunits respectively under conditions to favor attachment via lysines. The resulting protein is called "FlCRhR", short for fluorescein-labeled catalytic, rhodamine-labeled regulatory.

When properly prepared, the labeled holoenzyme (55,56) is found to have essentially identical sensitivity to cAMP as that of native holoenzyme. This property is valuable because it guarantees that the sensor has the optimal affinity for cAMP, since native holoenzyme is the main endogenous sensor for cAMP and the physiological concentrations of interest are precisely those which affect this enzyme. The ratio between fluorescein and rhodamine emissions increases by up to twofold upon cAMP-induced dissociation of the subunits. Moreover the activity of labeled enzyme as a kinase upon stimulation by cAMP is found to be identical to native holoenzyme likewise stimulated. This property is valuable because it minimizes any perturbation of the biological effectiveness of cAMP within cells into which the sensor has been introduced. If the sensor were not an effective kinase, then any cAMP bound to it would be prevented from exerting biological activity, so that the sensor would inherently inhibit the signaling pathway whose activity was to be measured. However, our sensor is a normal kinase, so cAMP molecules bound to it are biologically fully effective and not wasted. A final bonus of using the natural kinase as a starting point is that once the subunits have been dissociated, their fluorescein and rhodamine tags permit tracking their separate fates, which turn out to be of considerable additional interest because the free catalytic subunit is able to enter the nucleus (55) and phosphorylate transcription factors there.

So far FlCRhR has been microinjected and imaged in REF-52 fibroblasts (55,57), BC3H-1 smooth muscle-derived cells (55), PC-12 cells, neonatal cardiac myocytes, primary and transformed osteoblasts (58), chick ciliary ganglion neurons (59), rat Schwann cells, MA-10 Leydig tumor cells (56), LLC-PK$_1$ kidney cells, *Xenopus* oocytes, *Drosophila* embryos, fish melanocytes (60), and *Aplysia* sensory neurons both in culture and in excised ganglia (61). Space permits only the following single example of biological application. *Aplysia* sensory neurons, one of the classic model systems for analyzing neuronal plasticity (62), have been studied (61) in collaboration with Binyamin Hochner, Bong-kyun Kaang, and Eric Kandel of Columbia University. Confocal microscopy allowed separate observation of the nucleus, the surrounding cytoplasm of the cell body, and the peripheral processes. Bath application of the relevant neurotransmitter, 5-hydroxytryptamine, produced rapid increases in cAMP with remarkable spatial gradients, high in the processes, only slightly elevated in the cell body. These gradients did not seem to be due to gross barriers to cAMP diffusion, since cAMP directly microinjected into the cell body was seen to spread into the processes by simple diffusion with a diffusion constant just below 10^{-5} cm^{2}s^{-1}, close to the value in free solution. Thus cAMP signals in this system seem to spread in a relatively simple manner, without either the tremendous buffering and sequestration or the regenerative amplification that Ca^{2+} often shows. Optical sections through the nucleus showed that it tended to exclude the holoenzyme (injected into the cytoplasm) as long as cAMP remained at basal levels. Prolonged elevation of cAMP and dissociation of the holoenzyme caused gradual translocation of the catalytic subunit into the nucleus over tens of minutes. Translocation was reversible if cAMP was removed. The peak cAMP level reached by a variety of stimulation conditions was fairly well correlated with the subsequent extent of nuclear translocation. The observed gradient puts high cAMP where it is most needed for short-term plasticity, at the distal processes where the presynaptic terminals would be *in vivo*. Only strong or repeated stimulations would be able to raise cAMP in the cell body sufficiently to release the catalytic subunit to diffuse into the nucleus, phosphorylate transcription factors, and cause longer-term changes in gene expression (61).

Prospects for Other Complex Analytes

In principle, fluorescence energy transfer measurements analogous to these studies with FlCRhR should be useful to monitor protein-protein and protein-DNA association/dissociation in a wide variety of systems. Such interactions, involving receptors, G protein subunits, tyrosine kinases, calmodulin, effector proteins, transcription factors, and promoter sequences, are of tremendous importance in biological signal transduction. It is also conceivable that fluorescent sensors for other analytes such as cGMP, diacylglycerol, or inositol trisphosphate, which cause a marked conformational change in their native sensors but do not cause overt dissociation, might be engineered by putting the two fluorescent tags on the domains of the protein that undergo the largest change in mutual proximity. The particular virtue of fluorescence methodology is that it can be applied in intact cells with high spatial and temporal resolution while the cells are performing physiological functions. Although an organic chemical purist might scorn fluorescence chemosensors created by modifying natural proteins, biochemists and cell biologists have such a need for good fluorescent chemosensors that all approaches ranging from total organic synthesis to molecular biology are welcomed.

Acknowledgments

The work reviewed here from the author's laboratory was supported by the Howard Hughes Medical Institute and by NIH grants GM31004, EY04372, and NS27177.

Literature Cited

1. Tsien, R. Y., *Annu.Rev.Neurosci.* **12**, 227, 1989.
2. Tsien, R. Y., *Am.J.Physiol.* **263**, C723, 1992.
3. Barritt, G. J., "Communication within Animal Cells." Oxford University Press, New York, 1992.
4. Hardie, D. G., "Biochemical Messengers: Hormones, Neurotransmitters and Growth Factors." Champman & Hall, London, 1991.
5. Tsien, R. Y., *Annu.Rev.Biophys.Bioeng.* **12**, 94, 1983.
6. Tsien, R. Y., In "Optical Methods in Cell Physiology" (P. de Weer and B. Salzberg, Eds.), pp. 327-345. Wiley, New York, 1986.
7. Lakowicz, J. R., "Principles of Fluorescence Spectroscopy." Plenum Press, New York, 1983.
8. Uster, P. S. and Pagano, R. E., *J.Cell Biol.* **103**, 1221, 1986.
9. Herman, B., In "Fluorescence Microscopy of Living Cells in Culture Part B, Methods in Cell Biology vol. 30" (D. L. Taylor and Y. -L. Wang, Eds.), pp. 219-243. Academic Press, San Diego, 1989.
10. Jovin, T. M. and Arndt-Jovin, D. J., *Annu.Rev.Biophys.Biophys.Chem.* **18**, 271, 1989.
11. Dix, J. A. and Verkman, A. S., *Biophys.J.* **57**, 231, 1990.
12. Handbook of Biological Confocal Microscopy." Plenum Press, New York, 1990.
13. Minta, A. and Tsien, R. Y., *J.Biol.Chem.* **264**, 19449, 1989.
14. Tsien, R. Y. and Poenie, M., *Trends Biochem.Sci.* **11**, 450, 1986.
15. Bright, G. R., Fisher, G. W., Rogowska, J., and Taylor, D. L., In "Fluorescence Microscopy of Living Cells in Culture Part B, Methods in Cell Biology vol. 30" (D. L. Taylor and Y. -L. Wang, Eds.), pp. 157-192. Academic Press, San Diego, 1989.
16. Lakowicz, J. R., Szmacinski, H., Nowaczyk, K., and Johnson, M. L., *Cell Calcium* **13**, 131, 1992.
17. Tsien, R. Y. and Waggoner, A., In "Handbook of Biological Confocal Microscopy" (J. Pawley, Ed.), pp. 169-178. Plenum Press, New York, 1990.
18. Tsien, R. Y., *Nature* **290**, 527, 1981.
19. Tsien, R. Y., *Biochemistry* **19**, 2396, 1980.
20. Carafoli, E., *Annu.Rev.Biochem.* **56**, 395, 1987.
21. Tsien, R. Y., *Trends Neurosci.* **11**, 419, 1988.
22. Tsien, R. W. and Tsien, R. Y., *Annu.Rev.Cell Biol.* **6**, 715, 1990.
23. Campbell, A. K., "Intracellular Calcium." John Wiley and Sons, Chichester, 1983.
24. Gerig, J. T., Singh, P., Levy, L. A., and London, R. E., *J.Inorg.Biochem.* **31**, 113, 1987.
25. Pethig, R., Kuhn, M., Payne, R., Adler, E., Chen, T. -H., and Jaffe, L. F., *Cell Calcium* **10**, 491, 1989.
26. Grynkiewicz, G., Poenie, M., and Tsien, R. Y., *J.Biol.Chem.* **260**, 3440, 1985.
27. Vander Donckt, E., *Prog React Kinetics* **5**, 273, 1970.

28. Minta, A., Kao, J. P. Y., and Tsien, R. Y., *J.Biol.Chem.* **264**, 8171, 1989.
29. Kao, J. P. Y., Harootunian, A. T., and Tsien, R. Y., *J.Biol.Chem.* **264**, 8179, 1989.
30. Poenie, M., Tsien, R. Y., and Schmitt-Verhulst, A. -M., *EMBO J.* **6**, 2223, 1987.
31. Williams, D. A. and Fay, F. S., *Cell Calcium* **11**, 55, 1990.
32. Cuthbertson, K. S. R. and Cobbold, P. H., *Cell Calcium* **12**, 61, 1991.
33. Adams, S. R., Kao, J. P. Y., Grynkiewicz, G., Minta, A., and Tsien, R. Y., *J.Am.Chem.Soc.* **110**, 3212, 1988.
34. Adams, S. R., Kao, J. P. Y., and Tsien, R. Y., *J.Am.Chem.Soc.* **111**, 7957, 1989.
35. Ellis-Davies, G. C. R. and Kaplan, J. H., *J.Org.Chem.* **53**, 1966, 1988.
36. Kaplan, J. H., *Annu.Rev.Physiol.* **52**, 897, 1990.
37. Kaplan, J. H. and Ellis-Davies, G. C. R., *Proc.Natl.Acad.Sci.USA* **85**, 6571, 1988.
38. Harootunian, A. T., Kao, J. P. Y., Paranjape, S., and Tsien, R. Y., *Science* **251**, 75, 1991.
39. Adams, S. R. and Tsien, R. Y., *Annu.Rev.Physiol.* **55**, In press, 1992.
40. Gatto, V. J., Arnold, K. A., Viscariello, A. M., Miller, S. R., Morgan, C. R., and Gokel, G. W., *J.Org.Chem.* **51**, 5373, 1986.
41. Gandour, R. D., Fronczek, F. R., Gatto, V. J., Minganti, C., Schultz, R. A., White, B. D., Arnold, K. A., Mazzocchi, D., Miller, S. R., and Gokel, G. W., *J.Am.Chem.Soc.* **108**, 4078, 1986. 42. Harootunian, A. T., Kao, J. P. Y., Eckert, B. K., and Tsien, R. Y., *J.Biol.Chem.* **264**, 19458, 1989.
43. Negulescu, P. A., Harootunian, A. T., Tsien, R. Y., and Machen, T. E., *Cell Regulation* **1**, 259, 1990.
44. Sage, S. O., Rink, T. J., and Mahaut-Smith, M. P., *J.Physiol.* **441**, 559, 1991.
45. Lee, C. O. and Levi, A. J., *Ann.N.Y.Acad.Sci.* **639**, 408, 1991.
46. Jaffe, D. B., Johnston, D., Lasser-Ross, N., Lisman, J. E., Miyakawa, H., and Ross, W. N., *Nature* **357**, 244, 1992.
47. Lasser-Ross, N. and Ross, W. N., *Proc.R.Soc.Lond.B* **247**, 35, 1992.
48. Robison, G. A., Butcher, R. W., and Sutherland, E. W., "Cyclic AMP." Academic Press, New York, 1971.
49. Taylor, S. S., Buechler, J. A., and Yonemoto, W., *Annu.Rev.Biochem.* **59**, 971, 1990.
50. Terasaki, W. L. and Brooker, G., *J.Biol.Chem.* **252**, 1041, 1977.
51. Hayes, J. S. and Brunton, L. L., *J.Cyclic Nucleotide Res.* **8**, 1, 1982.
52. Murray, K. J., Reeves, M. L., and England, P. J., *Mol.Cell.Biochem.* **89**, 175, 1989.
53. Barsony, J. and Marx, S. J., *Proc.Natl.Acad.Sci.USA* **87**, 1188, 1990.
54. Deslongchamps, G., Galán, A., de Mendoza, J., and Rebek, J.,Jr., *Angew.Chemie Int.Ed.Engl.* **31**, 61, 1992.
55. Adams, S. R., Harootunian, A. T., Buechler, Y. J., Taylor, S. S., and Tsien, R. Y., *Nature* **349**, 694, 1991.
56. Adams, S. R., Bacskai, B. J., Taylor, S. S., and Tsien, R. Y., In "Fluorescent

Probes for Biological Activity of Living Cells - A Practical Guide" (W. T. Mason and G. Relf, Eds.), Academic Press, In press, 1992.
57. Harootunian, A. T., Adams, S. R., and Tsien, R. Y., *unknown* Manuscript in prepar, 1992.
58. Civitelli, R., Mahaut-Smith, M., Bacskai, B. J., Adams, S. R., Avioli, L. V., and Tsien, R. Y., *J.Biol.Chem.* submitted, 1993.
59. Gurantz, D., Harootunian, A. T., Tsien, R. Y., Dionne, V. E., and Margiotta, J. F., *Neuron* submitted, 1992.
60. Sammak, P. J., Adams, S. R., Harootunian, A. T., Schliwa, M., and Tsien, R. Y., *J.Cell Biol.* **117**, 57, 1992.
61. Bacskai, B. J., Hochner, B., Mahaut-Smith, M., Adams, S. R., Kaang, B. -K., Kandel, E. R., and Tsien, R. Y., *Science* **260**, 222, 1993.
62. Kandel, E. R., In "Principles of Neural Science" (E. R. Kandel, J. H. Schwartz, and T. M. Jessell, Eds.), Elsevier, New York, 1991.
63. Valet, G., Raffael, A., Moroder, L., Wünsch, E., and Ruhenstroth-Bauer, G., *Naturwissenschaften* **68**, 265, 1981.
64. Rink, T. J., Tsien, R. Y., and Pozzan, T., *J.Cell Biol.* **95**, 189, 1982.
65. Bassnett, S., Reinisch, L., and Beebe, D. C., *Am.J.Physiol.* **258**, C171, 1990.
66. Illsley, N. P. and Verkman, A. S., *Biochemistry* **28**, 1215, 1987.
67. Biwersi, J., Farah, N., Wang, Y. -X., Ketcham, R., and Verkman, A. S., *Am.J.Physiol.* **262**, C243, 1992.
68. Raju, B., Murphy, E., Levy, L. A., Hall, R. D., and London, R. E., *Am.J.Physiol.* **256**, C540, 1989.

RECEIVED June 21, 1993

Chapter 10

1,2-Bis(2-aminophenoxy)ethane-N,N,N',N',- tetraacetic Acid Conjugates Used To Measure Intracellular Ca^{2+} Concentration

Michael A. Kuhn

Molecular Probes, Inc., 4849 Pitchford Avenue, Eugene, OR 97402

We have developed several series of new indicators for intracellular free Ca^{2+} levels. These include a series of long wavelength fluorescent conjugates of BAPTA (1,2-(bis)-2-aminophenoxyethane-N,N,N',N'-tetraacetic acid) with extremely high quantum yields (up to 0.75), a series of dextran indicator conjugates for measurement of Ca^{2+} over long time periods and a series of Ca^{2+} indicators targeted to the lipid portion of cells for the measurement of near-membrane Ca^{2+} levels.

Fluorescent Ca^{2+} indicators are organic molecules in which reversible binding of a single free Ca^{2+} ion is accompanied by a change in excitation or emission. These compounds are fluorescent derivatives of the parent chelating compound BAPTA and are designed to report the level of free Ca^{2+} in the cytosol of cultured cells. Using recently developed imaging technology (1), fluorescent Ca^{2+} indicators are used to accurately visualize real time changes in cytoplasmic Ca^{2+} levels.

This approach to the imaging of intracellular Ca^{2+} flux was initiated by the work of Dr. Roger Tsien. At the University of California at Berkeley, he developed the Ca^{2+}-selective chelator, BAPTA (2) and it's initial fluorescent derivatives. With the invention of quin-2 in 1980 (3) and the subsequent indicators, fura-2, indo-1 (4), fluo-3 and rhod-2 (5), researchers in the biological sciences have begun to define the complex role of Ca^{2+} in living cells (6).

The Ca^{2+} selective chelator BAPTA is essentially an aromatic version of the widely used chelator EGTA (Figure 1). Replacing two alkyl bonds by aromatic rings lowers the pK_a of the ion-binding amines and makes Ca^{2+} binding insensitive to changes in pH over the physiological range. The aromatic ring is electronically conjugated to the binding site and changes in the electron density of the ring can alter the affinity for Ca^{2+}. For example, the affinity for Ca^{2+} decreases when the rings are substituted with electron withdrawing groups (nitro, bromo) and increases for electron donating groups (methyl) (1,7).

0097–6156/93/0538–0147$06.00/0

Masking the chelating carboxylic acid groups with acetoxymethyl (AM) esters, BAPTA and its fluorescent derivatives can be converted to nonpolar compounds that passively cross the plasma membrane. Once in the cell, non-specific esterases cleave the AM ester back to the free indicator and the resulting tetra or penta anion accumulates in the cytosol. Cells incubated in a very dilute dye solution (1-5 μM) still reach a useful concentration of indicator (>25 μM). This is an extremely valuable technique, especially when used with Ca^{2+} indicators that exhibit a shift in excitation or emission on ion binding. By these means, researchers can estimate intracellular Ca^{2+} changes in single cells or populations of cells (8).

Although this is an elegant system for the accurate determination of Ca^{2+} levels, variable properties of the cytosolic environment effect the dissociation constant of the indicator. These include the viscosity (9), ionic strength and pH (10) of the cytosol and the presence of proteins (11). In order to approximate intracellular ionic strength and pH, we report dissociation constants determined using solutions of 100 mM KCl and 10 mM MOPS buffer at pH 7.20. The concentration of free Ca^{2+} is buffered to the nanomolar range using 10 mM EGTA (12).

In addition to these considerations, dye concentration, incomplete hydrolysis of AM esters, photobleaching (13) and phototoxicity of the reporter, saturation of the binding site and transport of the indicator into vesicles all have an effect on the overall accuracy of the measurement. Some of these problems can be addressed by improving the properties of the ion indicator.

Indicator design Dyes designed for use as intracellular indicators of cytoplasmic Ca^{2+} have some fairly strict requirements that must be met before they can be considered for use in cells, including high selectivity for Ca^{2+} against background of millimolar levels of Na^+, K^+ and Mg^{2+}, bright fluorescence in water and insensitivity to pH changes near 7. It is also important for the fluorophore to be amenable to prodrug formation (AM ester) or at least not be highly lipophilic. These requirements present a challenge to the design of chemosensors for imaging of intracellular dynamics.

Synthesis of fluorescent Ca^{2+} indicators has revolved around incorporation of the aromatic BAPTA ring into conjugated heterocyclic systems. This type of synthetic approach has produced the dyes currently in use, but because of long, often low yield syntheses and the requirements of intracellular probes (e.g., pH insensitivity, high quantum yield), the number of indicators useful for imaging has been limited to mainly three dyes.

Fusing the BAPTA directly into a multi-ring system through a locked *trans*-ethylenic bond results in dyes that consistently exhibit a shift in excitation wavelength on Ca^{2+} binding (Figure 2). Using fura-2 in the cytoplasm, the ratio of the excitation peaks at 340 and 370 nm can be used to determine the fraction of the indicator that is bound to Ca^{2+}. This ratio is relatively insensitive to dye concentration, cell thickness and other artifacts.

Incorporation of the BAPTA ring into a fluorophore through a rotating *trans*-ethylenic bond gives an increase in fluorescence quantum yield (fluo-3) or a shift in emission wavelength on binding (indo-1) (Figure 3).

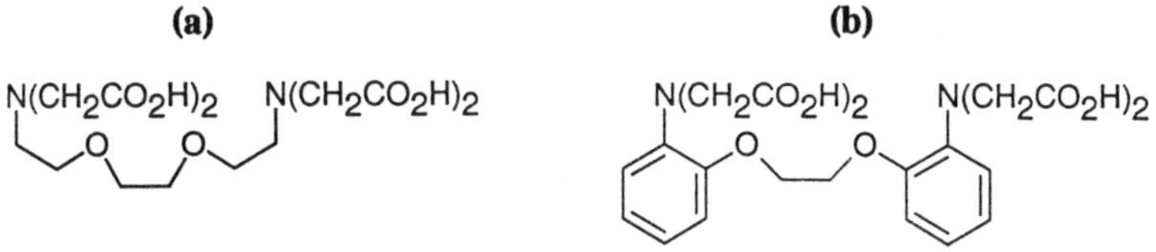

(a)   **(b)**

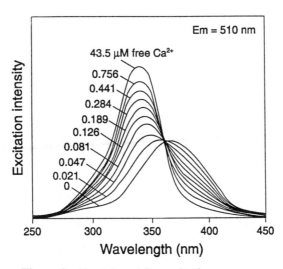

Figure 1. EGTA (a) and BAPTA (b)

Figure 2. Fura-2 and its excitation response

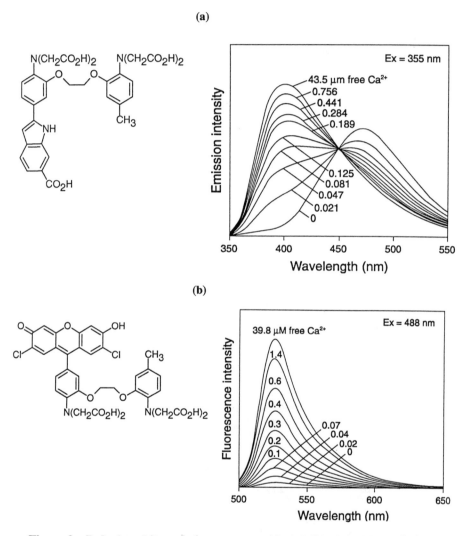

Figure 3. Indo-1 and its emission response (a) and fluo-3 and its emission response (b)

Quantum yield To measure free Ca^{2+} levels in a single cell it is important to load the dye at a low concentration because high intracellular concentrations can buffer cytoplasmic transients and give rise to fluorescence artifacts. This is best accomplished by achieving a strong emission signal from a small amount of dye. The signal that can be generated from a single dye molecule depends on the amount of light the molecule will absorb at the excitation wavelength (extinction coefficient) and the portion of that light that is re-emitted as fluorescence (quantum yield). Indicators that have low quantum yields (< 0.05) require high intracellular dye concentrations to achieve a useful signal.

As the fluorescence properties of Ca^{2+} indicators improve, the amount of dye necessary to achieve a meaningful signal decreases. The first of the fluorescent indicators for Ca^{2+} based on the BAPTA structure, quin-2, was loaded in cells as its AM ester at a concentration of about 30 μM. With the introduction of fura-2, the loading concentration dropped to about 1 μM, because the fluorescence intensity of fura-2 is approximately thirty times that of quin-2. Further improvement in emission intensity, whether due to increased extinction coefficient or quantum yield, would serve to further lower the amount of dye necessary to achieve a strong signal from a single cell.

There are relatively few that are highly fluorescent in water. The most widely used water soluble fluorophores are the fluorescein type. These dyes have extremely high quantum yields in water (up to 0.9) and would appear to be ideal for the basis of an indicator. Fluo-3 has a structure similar to fluorescein, but it is sufficiently different that it's quantum yield is only 1/8 that of the parent 2',7- dichlorofluorescein at it's brightest.

Excitation wavelength Light used to excite the intracellular indicator can also have harmful effects on the cell. High energy UV light (< 400 nm) is the most damaging and can cause cell death at high intensities. Use of dyes such as fura-2 or indo-1 requires ultraviolet excitation wavelengths (340 - 360 nm). Shifting the excitation maximum to longer wavelengths accommodates laser excitation sources and lowers cellular autofluorescence. Fluo-3 and rhod-2 were developed to provide indicators that are more compatible with lower energy Ar laser emissions (488 and 514 nm). These dyes have peak excitations wavelengths at 505 nm and 550 nm respectively. Extending the excitation wavelength even farther into the red would give an emission well removed from any cellular background fluorescence and further lower light damage to cellular structures.

Fluorescent Conjugates of BAPTA

We discovered that conjugation of high quantum yield fluorophores such as fluorescein and rhodamine to BAPTA through an amide linker yields indicators that exhibit an increase in fluorescence emission intensity on binding Ca^{2+}. Unlike previous indicators, these fluorescent conjugates of 5-amino BAPTA have more than one rotating sigma bond between the aromatic BAPTA binding site and the fluorescent "reporting" portion of the indicator. The long wavelength Ca^{2+} indicators, Calcium Green^{TM}-1, Calcium Orange^{TM} and Calcium Crimson^{TM}, are examples of fluorescent sensors based on this design (Figure 4).

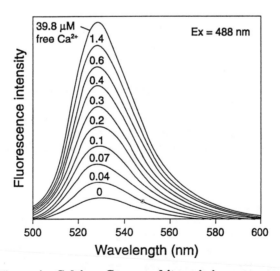

Figure 4. Calcium Green and its emission response

Attaching a fluorophore in the correct position relative to an ion binding site can result in a chemically stable, highly sensitive ion sensor where optical response is due to an electronic effect known as photon induced electron transfer (PET) (*14,15*). The electron density of the free binding site quenches the fluorescence of the covalently linked fluorophore. In the bound state some of the electron density of the indicator is shared with the bound species and the quenching effect is relieved, giving a large increase (up to 100-fold) in emission intensity of the fluorophore on interaction with a photon (Figure 5).

The heteroatomic spacer between the binding site and fluorescent reporter retains the desirable characteristics of the BAPTA Ca^{2+} binding site (pH insensitivity, high Ca^{2+} selectivity) along with the fluorescence properties of the attached fluorophore (quantum yield, emission wavelength, extinction coefficient). This synthetic method yields a series of dyes that can have a wide variety of ion binding affinities and emission wavelengths. For the purposes of intracellular imaging, we have focused on fluorophores that have excitation wavelengths greater than 488 nm, extinction coefficients of 70,000 and fluorescence quantum yields greater than 0.3. These fluorescent conjugates bind Ca^{2+} with a slightly lower affinity than the parent BAPTA compound, but changing fluorophores has little further effect on the dissociation constant. Because the fluorescent properties are retained and conjugation has little effect on the binding site, the properties of the resulting indicator can be predicted. These compounds represent a flexible new method for the design and synthesis of fluorescent ion indicators.

Calcium Green-1 has been shown to be less phototoxic than fluo-3 when used in sea urchin egg cells (*16*). Reported uses of Calcium Green include the imaging of spiral waves of Ca^{2+} in *Xenopus* oocytes (*17*) and to measure Ca^{2+} flux in conjunction with photoactivatable (caged) probes (*18*). Thermodynamic dissociation constants and dissociation rate constants for these indicators have been determined using stopped flow fluorescence measurements (*19*). The "Color Series" dyes were introduced in March 1991 (*20*). Since that time we have developed some derivatives that have slightly different responses:

Calcium Green-2 Conjugation of two fluorescein dyes to diamino BAPTA gives Calcium Green-2 (Figure 6), an indicator that has a greater dynamic range than the parent Calcium Green-1. When bound to Ca^{2+}, the indicator experiences a ~ 100-fold increase in fluorescence quantum yield to a maximum of 0.75, identical to Calcium Green-1.

In it's unbound form, the floppy ethylenedioxy bridge between the BAPTA rings allows the fluorophores to come close enough to quench each other. Confirmation of this comes from the absorption spectra of the Ca^{2+} bound and unbound indicator, the unbound dye has a broad absorbance, while the bound form shows a sharp peak at 505 nm. The bound form is identical to the parent Calcium Green-1, which undergoes no change in absorbance on ion binding.

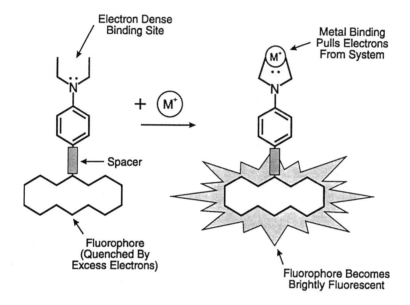

Figure 5. Photoinduced electron transfer (PET)

Table I: Data Tables

Indicator	K_dCa^{2+}	Ex High	Ex Low	Em High	Em High	ϵ max
Calcium Green-1	189 nM	506	506	534	533	88,000
Calcium Orange	328 nM	555	554	576	575	80,000
Calcium Crimson	205 nM	588	587	610	612	85,000

Indicator	QY High	QY Low	QY Increase
Calcium Green-1	0.75	0.06	12.5x
Calcium Orange	0.33	0.11	3.0x
Calcium Crimson	0.53	0.18	2.9x

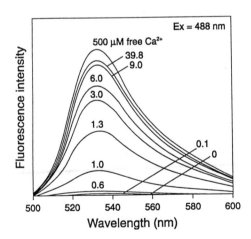

Figure 6. Calcium Green-2 and its emission response

The response of this indicator consists of two parts: a photoinduced electronic effect of Ca^{2+} binding that changes the electron density of the conjugated fluorophores and an "unquenching" effect caused by a physical separation of fluorophores. These each contribute to the improved dynamic range of response.

Additional evidence for the separation of fluorophores on binding comes from an X-ray crystallographic analysis of Ca^{2+}-bound difluoroBAPTA (21). This shows the aromatic positions *para* to the chelating nitrogens are held apart when BAPTA organizes around a single Ca^{2+}.

The addition of a second fluorescein dye decreases the Ca^{2+} affinity of Calcium Green-2 to about half that of Calcium Green-1. Because of it's increased dynamic range and slightly higher K_d, this maybe the dye of choice for some applications. It will be especially useful in cases where Ca^{2+} levels increase from resting levels (100-200 nM) to high concentrations (500 nM to >1 μM), such as in contracting muscle cells.

Calcium Green-5N   Based on the fluorophore-fluorophore quenching seen in Calcium Green-2, we designed a very low affinity Ca^{2+} indicator that uses a nitro group on the second BAPTA ring to quench the fluorescence of the reporting fluorophore when it is not bound to Ca^{2+} (Figure 7). The fluorophore-quencher interaction increases the dynamic range of response and the electron withdrawing effect of the nitro group lowers the affinity for Ca^{2+} from the nanomolar to the micromolar range.

Dextran Conjugates of Ca^{2+} Indicators

Since the commercial introduction of fura-2 by Molecular Probes, Inc. in 1985, problems with the indicator have been reported (22). The major difficulty is that the indicator does not remain in the cytoplasm for long periods of time. Within an hour of cell loading, a significant portion of the indicator typically localizes in intracellular compartments where it is still fluorescent, but is no longer sensitive to changes in cytoplasmic Ca^{2+} levels. The degree of compartmentation after depends on the type of cell and other factors, but generally limits the time during which useful measurements can be made. One approach to solving this problem is to covalently link the indicator to a high molecular weight, water-soluble carrier. We chose dextran as the carrier because of its transparency to light, biocompatibility and ease of use.

The preferred approach was to activate the free fura-2 and react it directly with an amino-substituted dextran. Since fura-2 has an "extra" carboxyl group on that is not involved in Ca^{2+} binding, attempts were made to block the chelating acid groups and couple the dye to the dextran by the remaining acid. Using Zn^{2+} to block the chelating site followed by reaction with amino dextran gave variable amounts of indicator substitution, which was largely Ca^{2+}-insensitive.

In a second attempt, the synthesis of fura-2 was modified to allow for the selective deprotection of the oxazole (non-chelating) carboxy. This method gave better success, but a direct comparison with free fura-2 showed that the quantum yield was only 15% to 20% that of the parent dye.

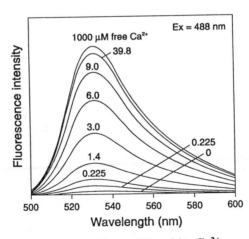

Figure 7.   Calcium Green 5N and its Ca²⁺ response

Ultimately the successful approach was to introduce a reactive site on the second BAPTA ring that could be conjugated to the dextran without effecting the fluorescence spectra. The resulting polymer conjugate has a response virtually identical to free fura-2 and a high quantum yield. This conjugate was tested through a collaboration with Dr. Peter Hepler and Dale Callaham of the University of Massachusetts Department of Botany. Their attempts to measure Ca^{2+} levels in plant cells (*lilium longiflorum*) using microinjected fura-2, potassium salt met with only limited success because the cells have efficient transport systems that remove the dye from the cytoplasm within minutes. When the dextran conjugate is microinjected into the cytoplasm of these cells, it remains sensitive to changes in Ca^{2+} for hours (*23*).

Since their introduction, the Ca^{2+}-indicating dextran conjugates have recently been used for a variety of applications such as neuronal tracing (*24*) and the ratiometric imaging of intracellular Ca^{2+} in conjunction with a Ca^{2+} insensitive long wavelength fluorescent dextran (*25*). In one interesting application, Calcium Green-1 dextran has been injected into zebrafish egg cells. After development, all the cells in the resulting embryo are labelled with the Ca^{2+}-sensitive dextran. Ca^{2+} changes throughout the body can then be imaged, including the beating of heart cells, using changes in emission intensity of the dye (Terasaki, M., NIH, personal communication, 1992).

Lipophilic Ca^{2+} Indicators

Since it became possible to conjugate Ca^{2+} indicators to almost any molecule or carrier, we became interested in directing the localization of the dyes to specific subcellular locations. One area of interest is the measurement of near-membrane Ca^{2+} levels.

The goal of our research in the area of lipophilic Ca^{2+} indicators is to develop fluorescent compounds that will remain near the surface of a membrane while reporting the Ca^{2+} concentration of the surrounding solution. By directing an indicator to intracellular membranes, one could measure Ca^{2+} dynamics in regions where currently available indicators will not work. These areas include not only the interior surface of the plasma membrane, but also the membranes of organelles such as the endoplasmic reticulum (ER), the Golgi and mitochondria. This type of indicator could be used for the measurement of ion levels in intracellular vesicles and neurons. Another possibility would be to combine lipophilic and hydrophilic indicators with two different fluorophores in a single cell and attempt to spatially distinguish Ca^{2+} gradients or waves in each domain.

Using BAPTA as the Ca^{2+}-binding portion of the molecule, three classes of fluorescent compounds have been designed that could be useful as indicators of near-membrane Ca^{2+}. The requirements of a near-membrane indicator are different from those of cytoplasmic indicators, which are designed to stay in the aqueous portion of the cell. For example, fura-2 has a polar carboxylate on the fluorophore to help prevent association of the dye with nonpolar structures.

Designing fluorescent Ca^{2+} indicators that will localize in the membrane allows for more flexibility in the choice of fluorophore than for an aqueous indicator. However, the extreme water solubility of BAPTA must be offset, either by the introduction of one or more lipophilic groups or by designing sensors that will distribute throughout the cytoplasm - but fluoresce only when near membranes. Ideally we can design a compound that can be passively loaded and that will localize to a specific region of the cell membrane after cleavage of the acetoxy methyl esters.

We have developed fluorescent indicators for Ca^{2+} that will partition into membranes, while still exhibiting a fluorescence response to Ca^{2+} binding. Development of these new indicators was approached from three different directions (in chronological order):

1. Lipophilic cyanine indicators
2. Long chain alkyl derivatives of currently available indicators
3. BAPTA derivatives with nonpolar or environment-sensitive fluorophores.

Merocyanine Indicators This type of indicator is a new conjugated fluorescent derivative of BAPTA that undergoes a large excitation shift on Ca^{2+} binding and has an emission wavelength of 710 nm. The structures of these indicators are based on asymmetrical cyanine dyes that have a dramatic increase in quantum yield on insertion into a lipid bilayer.

The dye tested was the fura analogue of the common fluorescent membrane probe "DiI" ($DiIC_{18}(3)$). Styryl dyes are not generally useful as fluorescent probes in aqueous solution because of their low quantum yield in water. However, when associated with lipids they become much more fluorescent. This is the case with $FuraIC_{18}$, which is approximately ten-times brighter when calibrated in the presence of liposomes than in buffer alone. The salt of this dye can be loaded into cells through a patch pipet and is seen to localize in the plasma membrane. This general structure can be easily modified and holds promise as a new class of membrane-associated indicators, for Ca^{2+} as well as other ions.

Alkyl Conjugates Using the same technology as for the synthesis of indicator dextran conjugates, single chain lipid conjugates were made from reactive derivatives of fura-2, Calcium Green-1 and indo-1. Lipophilic derivatives were tested for their response to Ca^{2+}, in aqueous solution and also when bound to liposomes. Comparison of these responses shows that insertion into the lipid layer is usually accompanied by an increase in fluorescence yield (up to 10-fold) and a decrease in affinity for Ca^{2+} ($K_d(aq) \approx 0.5 \mu M$; $K_d(lipid) \approx 1$ to $5 \mu M$). When loaded into the cell by patch-pipet, these dyes localize on the interior surface of the plasma membrane with no residual dye in the aqueous solution.

These dyes were tested in a collaboration with Dr. Fred Fay (University of Massachusetts), who has recently reported the use of C_{18} fura-2, potassium salt to image near-membrane Ca^{2+} transients by loading through a patch-pipet (26).

To increase the lipophilicity of the single chain probes, the phosphatidyl ethanolamine (PE) and cholesterol derivatives of Calcium Green-1 were synthesized. The AM derivative of Calcium Green-1 cholesterol shows no fluorescence after incubation for up to twelve hours with 3T3 cells, which readily cleave Calcium Green-1 AM. All the AM ester derivatives of these alkyl conjugates appear to be poor substrates for esterases, a problem that seems to get worse as the fatty portion of the dye increases. Loading the Calcium Green-1 cholesterol AM into 3T3 cells resulted in faint fluorescence that was associated with about 1% of the cells. It is likely that the lipophilic nature of the indicator traps it in the membrane and prevents the esters from coming into contact with cytoplasmic esterases. Unlike the cytoplasmic-targeted indicators, *in vitro* hydrolysis of the alkyl Calcium Green-1 AM derivatives with base initially gives a thick, colorless precipitate that is the detergent-like partially hydrolysed indicator. The solid eventually dissolves into a fluorescent solution when all the esters and acetates are cleaved.

Environment Sensitive Fluorophores To look more closely at the type of dyes that will load effectively as their AM ester, we synthesized other indicators containing fluorophores that fluoresce brightly only when in a nonpolar environment. Only two of these dyes have been tried: the pyrene and DANSYL (5-dimethylamino naphthalene-1-sulfonyl) derivatives. Although the DANSYL derivative is not very lipophilic, its fluorescence is highly environment sensitive (27).

Loading of the dipyrene BAPTA AM resulted in the same weak staining as seen with the alkyl conjugates. The diDANSYL BAPTA does load and cleave to give an indicating species that exhibits an emission increase near 570 nm on Ca^{2+} binding. The localization of this dye was not determined, but the staining pattern appears similar to fura-2 AM loading. This dye responds when the level of intracellular Ca^{2+} is increased using ionomycin.

The design and synthesis of new indicators for measuring free Ca^{2+} near the surface of intracellular membranes continues. We have seen membrane localization of fatty indicators that are introduced into cells by patch pipet, but loading of AM esters of this type of dye has proven difficult. The key to solving this part of the problem may lie in choice of the fluorophore.

The next generation of membrane-directed Ca^{2+} indicators will concentrate on structures that are readily cleaved by cytoplasmic esterases but localize in the membrane after cleavage. The best fluorophore will be one that is brightly fluorescent only in a lipid environment. Such a dye could cross the plasma membrane, be cleaved to the indicator in the aqueous portion of the cell and then fluoresce only when it is associated with nonpolar structures. It is expected that the levels of Ca^{2+} very close to the surface of the plasma membrane will be much higher than normal cytoplasmic levels. This will require further synthetic modification to tune the response to a useful range.

In summary, we have tried to address reported problems with the use of fluorescent indicators for Ca^{2+} measurement through the synthesis of new indicators and derivatives. We have increased the quantum yield and extinction coefficients to improve the fluorescence signal; increased the available excitation wavelength to reduce the excitation energy and increased the effective molecular weight of the indicator to prevent transport out of the cytoplasm. In addition, lipophilic indicators are under investigation as a means of targeting indicators to specific subcellular areas.

LITERATURE CITED

1. Gross, D.J. In *Noninvasive Techniques in Cell Biology*; Wiley-Liss, Inc. 1990, pp. 21-51
2. Tsien, R.Y. *Biochemistry* **1980**, *19*, 2396
3. Tsien, R.Y.; Pozzan, T.; Rink, T.J. *J. Cell Biol.* **1982**, *94*, 325
4. Grynkiewicz, G.; Poenie, M.; Tsien, R.Y. *J. Biol. Chem.* **1985**, *260*, 3440

5. Minta, A.; Kao, J.; Tsien, R.Y. *J. Biol. Chem.* **1989,** *264,* 8171
6. Tsien, R. In *Methods in Cell Biology*; Taylor, D.L. and Wang, Y.L. Eds. Academic Press: 1989 Vol. 30 pp 125-156
7. Pethig, R.; Kuhn, M.; Payne, R.; Adler, E.; Chen, T.-H; Jaffe, L.F. *Cell Calcium* **1989,** *10,* 491
8. Uto, A.; Arai, H.; Ogawa, Y. *Cell Calcium* **1991,** *12,* 29
9. Poenie, M. *Cell Calcium* **1990,** *11,* 85
10. Harrison, S.M.; Bers, D.M. *Biochim. Biophys. Acta* **1987,** *925,* 133
11. Roe, M.W.; LeMasters , J.J.; Herman, B. *Cell Calcium* **1990,** *11,* 63
12. Tsien, R.; Pozzan, T. *Meth. Enzymol.* **1989,** *127,* 230
13. Becker, P. et. al. *Am. J. Physiol.* **1987,** *253* C3
14. de Silva, A.P.; Sandanayake, K.R.A. *Tet. Lett.* **1991,** *32,* no.3, 421
15. Bryan, A.J.; de Silva, A.P., de Silva, S.A., Rupasinghe, R.A.D.D.; Sandanayake, K.R.A.S. *Biosensors* **1989,** *4,* 169
16. Stricker, S.A.; Centonze, V.E.; Paddock, S.W.; Schatten, G. *Dev. Biol.* **1992,** *149,* 370
17. Girard, S.; Luckoff, J.L., Sneyd, J.; Clapham, D *Biophys. J.* **1992,** *61,* 509
18. Bird, G.S.J; Obie, J.F.; Putney, J.W. *J. Biol. Chem.* **1992,** *267,* 17722
19. Eberhard, M.; Erne, P. *Biochem. Biophys. Res. Comm.* **1991,** *180,* 209
20. Kuhn, M.A. *Bioprobes, Molecular Probes, Inc.* **1991,** *13,* 1
21. Gehrig, J. et. al. *J. Inorg. Biochem.* **1987,** *31,* 313
22. Moore, E.D.W.; Becker, P.L.; Fogarty, K.E.; Williams, D.A.; Fay, F.S. *Cell Calcium* **1990,** *11,* 157
23. Miller, D.D.; Callaham, D.A.; Gross, D.J., Hepler, P.K. *J. Cell Science* **1992,** *101,* 7
24. O'Malley, D.M.; Lu, S.M.; Guido, W.; Adams, P.R. *Neuroscience* **1992,** *18* 142
25. Gilroy, S.; Jones, R.L. *Proc. Nat. Acad. Sci USA* **1992,** *89,* 3591
26. Etter, E.F.; Kuhn, M.A.; Fay, F.S. *Biophys. J.* **1993** *64* A364
27. Haugland, R.P. *Handbook of Fluorescent Probes and Research Chemicals, Molecular Probes, Inc.* **1992,** 113

RECEIVED April 19, 1993

Chapter 11

Fluorescent Chemosensors for Monitoring Potassium in Blood and across Biological Membranes

Divakar Masilamani and Mariann E. Lucas

Biotechnology Department, Allied-Signal Inc., Morristown, NJ 07962-1021

A rational approach to developing a fluorescent chemosensor for potassium (K^+) is presented. In this approach, J-M. Lehn's [222] cryptand, which selectively binds K^+, is covalently attached to coumarin at positions 6 and 7. In this hybrid system, coumarin plays the role of a transducer translating the free energy of supramolecular interaction between the cryptand and K^+ into measurable enhancement of its fluorescence. By immobilizing this system on an optical fiber, the continuous monitoring of K^+ in patients undergoing open-heart surgery can be achieved. In methanol, the hybrid system behaves as a reagent and can be used in automated microfluorometers to assay potassium in the range of 0-6 mM in the presence of 500-3000 fold excess of sodium (Na^+) with 99% accuracy. The fluorescent chemosensor also serves as a tool for studying rates of transport of K^+ across biological membranes. A mechanism based on photo-induced electron transfer explains the observed fluorescence enhancement caused by K^+ binding.

In this paper, we report our success in developing a practical fluorescent chemosensor for K^+ (1). This sensor was specifically designed for attachment to an optical fiber for the continuous sensing of K^+ in the extracorporeal blood of patients undergoing open-heart surgery. At the present time, K^+ in blood is monitored through a batch process. Blood samples are withdrawn periodically from patients and sent for analysis by conventional techniques such as atomic absorption spectrometry (2). There is a

0097-6156/93/0538-0162$06.25/0

lag-time between sampling and reporting of the results which can be critical. A sudden surge in the level of K^+ (above 6 mM) may be an indication that the patient is going into "shock". The lag-time can be avoided through continuous monitoring. Our fiber optics-based sensor will provide this technology.

A rational approach to developing a fluorescent chemosensor requires an understanding of the subtle differences between classical reagents and sensors (3). The critical thermodynamic factor that distinguishes the sensor from the reagent can also be used to convert one to the other. The thermodynamic characteristics of sensors clearly indicate a need for efficient transducers (4) that translate the weak chemical free energy of sensor-analyte interactions into measurable changes in their physical properties. We chose fluorophores for this purpose not only because they are highly sensitive (5) to electronic perturbations, but are also ideal for remote sensing through optical fibers (6). It is important that the design of our sensor takes into account the conditions that prevail in the extracorporeal blood. The relative concentrations of other alkali metal ions, their diameters and solvation energies, etc. will define the selectivity limits that are required of the sensor for K^+.

The design of fluorescent chemosensors for K^+ is simple in its logic. It combines known ionophores that are highly selective in binding K^+ with a fluorescent group. The two units are combined in such a way that the supramolecular interactions of K^+ that cause the electronic perturbations of the ligating atoms of the ionophore will be transmitted to the fluorescing chromophore through conjugated double bonds. The challenge is not in designing the hybrid sensors. It is in making the right choice of the components that constitute the hybrid. Here chemical intuition plays a significant role.

Once the decision was made on the fluorescent signaling group, several hybrid molecules incorporating the signal group in several K^+-selective ionophores were designed and synthesized. The best candidate sensor for K^+ was identified. The synthesis represents the most frustrating and time-consuming part of our research. This aspect will not be discussed here. However, an outline of our synthesis is presented as Appendix I at the end of the chapter and details are provided in two U. S. Patents (1).

Fluorescent Sensors and Reagents

In developing a fluorescent sensor for K^+, we need to understand the subtle differences between a classical reagent and a sensor (3,4).

Historically, reagents were designed to consume and transform analytes. Reagents, therefore, are not reusable. They are suited for batch analysis in which the transformed analyte is easier to analyze and quantitate. There is a time-lag between sampling and reporting of the result. This approach is expensive in terms of time, material and labor.

Sensors on the other hand, are devices that continuously monitor analytes (4,6). Sensors do not consume analytes. They interact with them in a reversible fashion. Reversibility implies lower equilibrium constants for interactions between the sensor and the analyte. Consequently, the free energy changes for these interactions are also small. Sensors, therefore, need efficient transducers that amplify their weak chemical interaction energies into strong measurable signals (4,6,7). We chose fluorophores as transducers for our K^+ sensors. There are several advantages to this approach. The measurement of fluorescence is an extremely sensitive technique (5). A number of efficient fluorescent systems such as dyes, fused aromatics, etc. with quantum yield above 0.9 are readily available. More importantly, fluorescent sensors can be immobilized on optical fibers in remote in vivo monitoring (6) of K^+ in the blood of patients undergoing open-heart surgery.

There is, however, a disadvantage which is common to all sensors. Sensors obey mass law since the equilibrium constants of their interaction with analytes are much lower. As a consequence, a plot of the sensor signal versus log of concentration of analyte is "S" shaped and not linear (6). This implies that the concentration of analyte (in our case K^+) must fall within the "dynamic range" of the "S" curve (6). Shifting the "dynamic range" will require a fundamental change in the way the sensor is constructed.

There is also an advantage resulting from the mass law effect. Lower concentration of a given sensor can scan large concentration ranges of the analyte. This advantage, along with those discussed earlier, supports our conclusion that sensors in general are economical to use. They are not consumed and are needed in small amounts. They respond instantaneously to changes in analyte concentration. In addition, fluorescent sensors are suited for remote sensing, miniaturization and automation (6).

Ions in Human Blood

In designing a fluorescent sensor for the in vivo monitoring of K^+, it is important to take into account the nature, composition and the environment surrounding the ions that are present in human blood. Table I provides the

Table I. Ions in Blood Serum

Ions	Normal Range (mmol/L)
Na^+	135-148
K^+	3.5-5.3
Ca^{+2}	4.5-5.5
HCO_3^-	23-30
Cl^-	103
Li^+	0-2.0

pH Range 7.35 - 7.45

Table II. Ionic Diameters and Free
Energy of Hydration

Cation	Ionic Diameter (Å)	$-\Delta G°$ (Kcal/mol.)(25°)
Li^+	1.20	122.0
Na^+	1.90	98.5
K^+	2.66	80.5
Ca^{+2}	2.12	379.0

essential data (Saari, L.A., Instrumentation Lab Inc., personal communication, 1984.) The medium is aqueous (near neutral pH). Na^+ is present in 35-40 fold excess compared to K^+. Because of its abundance and its similar chemical behavior, Na^+ will compete with K^+ for the sensor. In the face of such a competition, K^+ can be analyzed reliably only if the sensor shows a selectivity of the order of 500 for K^+ over Na^+. This is a formidable task.

Even though Ca^{2+} is present in amounts comparable to K^+, the doubly-charged metal ion is easy to discriminate against. Grell et al. (8) have shown that divalent calcium is complexed better by anionic carriers rather than the neutral ones.

Overcoming the large free energy of hydration (9) (Table II) of alkali and alkaline earth metal ions in aqueous medium is also a challenging task. The sensor designed must compete with water for the designated ion. Fortunately, the hydration energy for K^+ is much smaller than that for Na^+ because of its larger size. Yet, is large enough to compete with the sensor for K^+.

Since our fluorescent sensor will be immobilized on an optical fiber, it is important to evaluate the overall performance of such a device. In general, optical fiber technology has given optical sensors an edge over other conventional technologies such as electrochemical probes. Optical sensors are not subject to electrical interferences and do not require a reference. The biocompatibility of optical fibers for in vivo applications has been well established (6). As discussed earlier, these devices are amenable to miniaturization (6). However, there are also disadvantages. Ambient light often interferes. Photo-bleaching of the sensor molecule may affect its long-term stability (6).

Design of Fluorescent Chemosensors for K^+

As outlined earlier, our fluorescent sensors for K^+ are hybrid systems made by attaching ionophores (of high selectivity for K^+) to efficient fluorescing groups. In designing these hybrid systems, it is important not to compromise either the selectivity of the ionophore or the efficiency of the fluorescing function. For this reason, we decided not to use natural antibiotics such as valinomycin in our hybrid systems. Valinomycin shows an unusually high selectivity for K^+ (9,10). However, the selectivity of this antibiotic is controlled by the way it folds to accommodate K⁺ (9). Conformational factors are therefore important. Any external attachment to this antibiotic will affect the folding mechanism thus affecting its selectivity.

On the other hand, synthetic ionophores are better suited for attachment to fluorophores. Their ion selectivities are related to the size of their cavities (9,10) which is usually unaffected by attachments to other molecules. Further, conformation is not an overriding factor. The fusion of a fluorescing group to synthetic ionophores will not significantly change their selectivities.

Table III summarizes the efficiencies of selected crown ethers for binding Na⁺ and K⁺ (11). Efficiencies are measured in terms of log of association constants (Ka) in methanol and water. The correlation between the diameter of the alkali ions and the cavity diameter of the crown ethers is evident from the enhanced association constants. For example, 18-crown-6, (cavity diameter 2.6-3.2 Å) binds K⁺ (diameter 2.66 Å) more selectively than Na⁺ (diameter 1.9 Å). On the other hand, 15-crown-5 (cavity diameter 1.7-2.2 Å) is more selective for Na⁺ than K⁺. The case of 21-crown-7 is interesting. Gokel et al. (12) have shown that the cavity of this ionophore (diameter 3.4-4.3 Å) is too large for K⁺ but larger still for Na⁺. As a result, it is more selective for K⁺ than 18-crown-6. The difference in log Ka values in methanol between K⁺ and Na⁺ for 21-crown-7 is larger (1.9) compared to 18-crown-6, for which the difference is 1.78. However, 21-crown-7 is less sensitive than 18-crown-6 in binding K⁺ (log Ka of 4.35 versus 6.1).

In water, log Ka values are 2-4 orders of magnitude smaller. Because of the high solvation energies of K⁺ and Na⁺ with water, it is difficult to extract them into the ionophore. The lower log Ka values imply that the ions are exchanged faster in water than in methanol. Consequently, the ionophores will behave more as sensors in water than in methanol.

Cryptands are basket-like bicyclic ionophores in which three strands of polyethers are tied together by two nitrogen atoms. They provide three-dimensional spaces for binding metal ions (11,13,14). They are several orders of magnitude more selective than crown ethers in binding alkali metal ions. Table IV provides data on dimensions and log Ka values for K⁺, Na⁺ and Li⁺ in water for [222] [221] and [211] cryptands. For the [222] cryptand, the difference in the log Ka value between K⁺ and Na⁺ in water is 2.54. For 18-crown-6, the difference (Table III) is 1.83.

Even though the ion-recognizing capabilities of the crown ethers and cryptands are well established, they are still "passive" sensors. The conversion of these "passive ionophores" into efficient fluorescent sensors depends on the choice of fluorogenic group and the way it is attached to these ionophores.

Selecting the fluorogenic group was challenging. We initiated our quest with naphthalene. 2,3-Naphtho-18-crown-6 was synthesized and evaluated. This particular ionophore has been shown by Pedersen (15) and Cram et al. (16) (through the extraction technique) to be highly selective for K^+ over Na^+. However, it was insoluble in water. In dioxane, a slight quenching of its fluorescence was observed on adding KI and NaI. K^+ was more selective than Na^+. The poor response of the fluorescence of this ionophore to metal ion- binding is probably due to the weak dipole generated in its excited state. The charges tend to be delocalized in the naphthalene systems. We need a system in which the excited state dipole moment is strong and localized. We looked for a system with heteroatoms. Coumarin is one such system (17,18). The fluorescence quantum yield is of the order of 0.9 for coumarin as well as its 7-hydroxy and 6,7-dihydroxy compounds (19). The Stokes shift is close to 100 nm. Even more importantly, coumarins absorb light between 300-400 nm which is a suitable range for use in optical fibers. In addition, coumarins are more soluble in water than fused aromatics such as naphthalene and anthracene (20).

Figure 1 shows the structure of coumarin and the numbering system. It also shows the ground state and the dipolar excited state of the 7-hydroxy and 6,7-dihydroxycoumarins (18). The reported fluorescence quantum yields of these two compounds are high (18,19). As discussed earlier, the way the fluorescing signal group is attached to the ionophore is of great importance. The efficiency of transduction of the free energy of supramolecular interaction (between the metal ion and ligating heteroatoms) depends of the ease of the electron transport from the heteroatoms to the fluorescing chromophore. In our design, the oxygen at position 7 of the coumarin will play a critical role in such electron transports. This oxygen atom will convert the "passive ionophore" into ion-recognizing chemosensor. We use the term chemosensor to emphasize the totally synthetic nature of our sensor as opposed to the natural or semi-natural systems which depend at least in part on nature to do the design and synthesis.

Materials and Methods

N-[2-Hydroxyethyl]piperazine-N'-[2-ethanesulfonic acid] (HEPES) was obtained from Sigma. NH_4Cl, $CaCl_2$, $MgCl_2$, LiCl, NaCl, KCl, ethanol and methanol were purchased from Fisher. DMSO and quinine sulfate were procured from Aldrich. The water used was purified by MILLI-Q water system (Millipore). Stock solutions (2 mM) of the 4-methylcoumaro-cryptands were prepared in DMSO and diluted with water or methanol. Stock solutions (0.1 M) were prepared in 50:50 ethanol-water for 4-methylcoumaro-crown ethers and diluted with the same solvent mixture.

Table III. Crown Ethers

	Cavity Diameter	log (Ka)$_{K+}$ (2.66Å)	log (Ka)$_{Na+}$ (1.90Å)
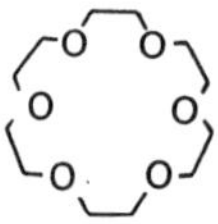	2.6-3.2Å	6.10 (M) 2.03 (W)	4.32 (M) 0.80 (W)
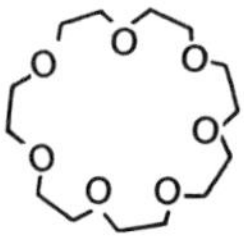	3.4-4.3Å	4.35 (M)	2.45 (M)
	1.7-2.2Å	2.21 (M) 0.76 (W)	3.24 (M) 0.79 (W)

Table IV. Cryptands

	Cavity Diameter	log (Ka)$_{K+}$ (2.66Å)	log (Ka)$_{Na+}$ (1.90Å)	log (Ka)$_{Li+}$ (1.20Å)
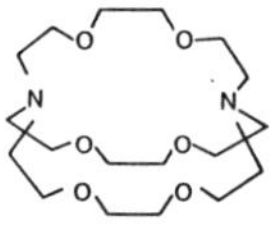	2.8Å	9.74 (W)	7.20 (W)	1.25 (W)
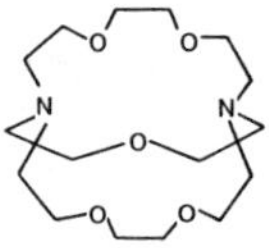	2.2Å	3.95 (W)	5.40 (W)	2.50 (W)
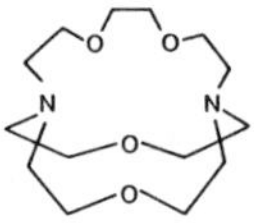	1.6Å	2.0 (W)	3.20 (W)	5.5 (W)

(W)Water   (M)Methanol

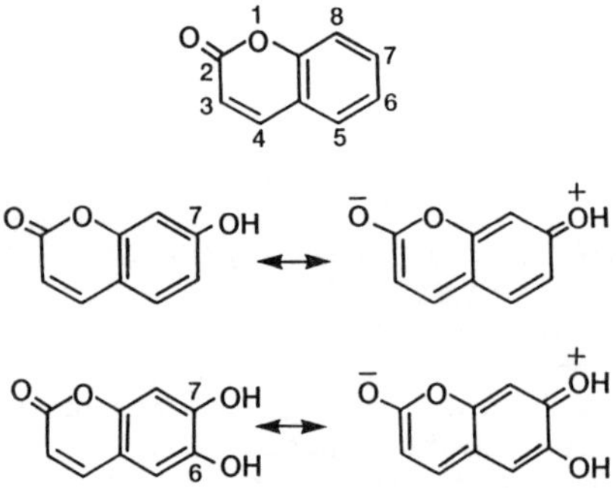

Figure 1   Coumarins

Spectral measurements were performed on a Perkin-Elmer LS-5 fluorometer using 5-nm slits. Relative quantum yield measurements were determined on a Perkin-Elmer LS-100 fluorometer. All studies were performed at 22°. The spectra were not corrected for the spectral variation in the excitation source and photomultiplier tube sensitivity. Background scattered light was zeroed electronically before the acquisition of the fluorescence of 4-methylcoumaro-[222]cryptand. Absorption spectra were obtained using Perkin Elmer Lambda 7 UV/VIS Spectrophotometer.

The fast atom bombardment mass spectrometer (FABMS) was a VG Analytical 2AB-HF reverse geometry double focusing instrument with an ion source using 7 KEV Xenon. Spectra were recorded on a VG 11-250 instrument control and an ion energy of 8 KEV. Positive ion spectra were more intense and generally of great analytical value. Competitive alkali ion binding with the 4-methylcoumaro-cryptands were conducted in glycerol matrix containing KCl and NaCl.

Results and Discussion

Evaluating the Feasibility of Using 4-Methylcoumarin in Chemosensors for K$^+$. We synthesized two classes of chemosensors: 4-methylcoumaro-crown ethers and 4-methylcoumaro-cryptands. These are shown in Figures 2 and 3. The feature that is common to all these chemosensors is the bridging of the ionophores to the coumarin through oxygens at position 6 and 7. We have already discussed the important role assigned to the oxygen at position 7. It may also be appropriate to point out that the nitrogens of the cryptand series play a role in "switching on" fluorescence. The crown ether series which do not contain nitrogens tend to "switch off" fluorescence (Vide infra). The methyl group in these chemosensors will be used later for attaching them to optical fiber.

In developing a new technology for fluorescent chemosensors, it is important to establish that the observed changes in the fluorescence emission truly reflect the anticipated behavior of their ionophore component. The ionophore components in the six chemosensors (Figures 2 and 3) have been discussed earlier. Their log Ka values for binding K$^+$, Na$^+$ and Li$^+$ have been listed in Tables III and IV.

The feasibility studies were first carried out with the 4-methylcoumaro-crown ethers. A 10^{-5} M solution of 4-methylcoumaro-18-crown-6 in 50:50 ethanol-water was excited at 330 nm. The fluorescence emission was observed at 410 nm. A pronounced quenching of the fluorescence was observed with addition of KCl. The quenching was less pronounced for NaCl. LiCl and CaCl$_2$ showed no quenching at all. A plot of Io/I (the ratio of initial fluorescence intensity to the intensity of the quenched fluorescence)

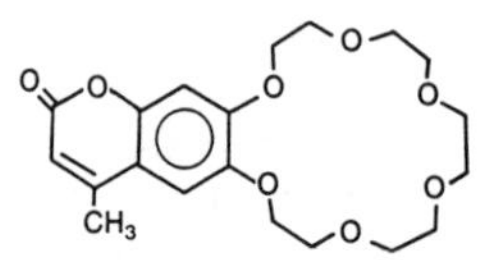

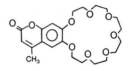

4 – (Methyl) Coumaro – 18 – Crown – 6   4 – (Methyl) Coumaro – 21 – Crown – 7

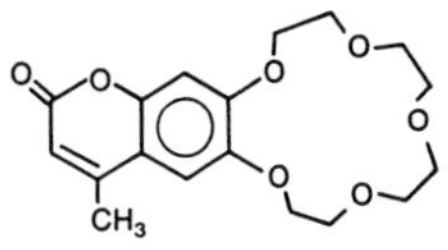

4 – (Methyl) Coumaro – 15 – Crown – 5

Figure 2   4-Methylcoumaro-crown ethers

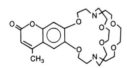   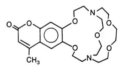

4 – (Methyl) Coumaro (222) Cryptand   4 – (Methyl) Coumaro (221) Cryptand

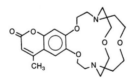

4 – (Methyl) Coumaro (211) Cryptand

Figure 3   4-Methylcoumaro-cryptands

against the concentration of the added salts (KCl and NaCl) is shown in Figure 4. The anticipated selectivity of this chemosensor for K^+ is clearly demonstrated.

We were particularly interested in the behavior of 4-methylcoumaro-21-crown-7. We have discussed earlier that the parent 21-crown-7 is more selective for K^+ over Na^+ than 18-crown-6 *(12)*. The latter however was more sensitive in its response to K^+ *(12)*. This is clearly demonstrated in Figure 4 which also includes a plot of Io/I against concentration of the added salts for the 4-methylcoumaro-21-crown-7. The ratio of Io/I for K^+ and Na^+ for this chemosensor is much larger than the corresponding ratio for the 4-methylcoumaro-18-crown-6, indicating higher selectivity of the former system for K^+. However, the 4-methylcoumaro-18-crown-6 is definitely more sensitive to K^+ binding. These results demonstrate beyond doubt that the 4-methylcoumarin is indeed an excellent fluorescent transducer that accurately reflects the ion selectivities and sensitivities of the ionophores to which it is conjugated.

Even though the feasibility of using coumarin in chemosensors has been established with the crown ethers, these systems are not sufficiently sensitive to be used in commercial devices. Since cryptands are several orders of magnitude more selective and sensitive than crown ethers *(9,13,14)*, we synthesized three 4-methylcoumaro-cryptands shown in Figure 3 and evaluated their performance.

4-Methylcoumaro-[222] Cryptand as Fluorescent Chemosensor for K^+. All three cryptand-based chemosensors shown in Figure 3 performed far beyond our expectation. Unlike the crown ether systems which showed selective fluorescence quenching, the cryptands showed selective fluorescence enhancement. A 4 μM solution of the 4-methylcoumaro-[222] cryptand (MCC222) in water (pH 7.4) showed a three-fold enhancement of its fluorescence emission when treated with 20 mM KCl. The addition of Na^+, Mg^{2+}, Ca^{2+} and NH_4^+ did not change its fluorescence. Similarly, 4-methylcoumaro-[221]cryptand and 4-methylcoumaro[211]cryptand showed selective fluorescence enhancement for Na^+ and Li^+, respectively. The latter two compounds however will not be discussed in this chapter.

Figures 5a and 5b show the fluorescent excitation and emission spectra of MCC222 at varying concentrations (0-20 mM) of KCl in an aqueous medium containing HEPES (pH 7.4). Similar results were obtained in 50:50 ethanol-water as well as in methanol. The excitation and emission peaks were at 340 nm and 420 nm, respectively. There is no spectral shift either in the excitation or emission spectra. The dose-response curve for concentrations ranging from 0-10 mM of KCl is shown in Figure 6. The range of concentration of K^+ in human blood is between 3.5-5.3 mM. This falls beautifully within the dynamic range of MCC222.

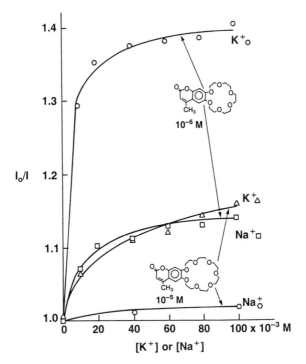

Figure 4 Selectivities   and   Sensitivities   of   4-
Methylcoumaro-crown ethers towards K⁺ and Na⁺

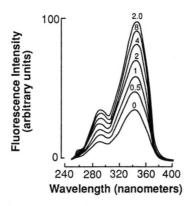

Figure 5a Excitation   spectra   of   4 μM   4-Methylcoumaro
[222] Cryptand in aqueous solution containing
5 mM HEPES (pH 7.4) and 0-20 mM KCl. Emission
420 nm (5 nm slits).

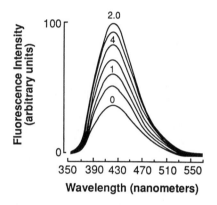

Figure 5b Emission spectra of the aqueous solutions described above. Excitation 340 nm (5 nm slits).

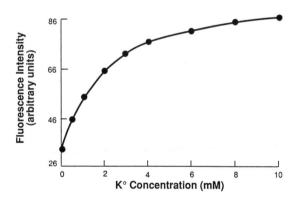

Figure 6 Calibration of 4 μM 4-Methylcoumaro[222] Cryptand in Aqueous Solution (pH 7.4)

The properties of MCC222 are summarized in Table V. The log Ka values were measured from the changes in the fluorescence. The log Ka in methanol was measured in the presence of 1.4 mM NaCl. This value is roughly two orders of magnitude larger than that observed in water. K^+ is bound much stronger to MCC222 in methanol.

The response of MCC222 for other ions was investigated in a systematic manner by two sets of experiments. First, a 4 μM solution of MCC222 was treated incrementally with Na^+ (ranging from 0-140 mM), Ca^{2+} (ranging from 0-10 mM) and NH_4^+ (ranging from 0-10 mM). No change in the fluorescence emission was observed for these salts. In the second series of experiments, a 4 μM solution of MCC222 was first treated with 4 mM KCl, which caused an instantaneous enhancement of the fluorescence and then remained constant at that level. The Na^+, Ca^{2+} and NH_4^+ salts were added in incremental amounts as before. Ca^{2+} and NH_4^+ showed no further change in the fluorescence. However, the addition of Na^+ decreased the fluorescence incrementally without altering the spectral properties of the probe. This suggests that Na^+ can displace K^+ from its binding site, yet the bound Na^+ does not show any change in the fluorescence of the chemosensor. The selectivity of the parent [222] cryptand for K^+ over Na^+ has been reported to be 350 (14) corresponding to a difference in log Ka of 2.54. Our chemosensor is more rigid because of the fusion of the aromatic ring of coumarin to the [222] cryptand. The selectivity is expected to be lower for K^+. The log Ka for Na^+ cannot be determined fluorometrically since Na^+ does not affect the fluorescence of this chemosensor in the absence of K^+. However, we can determine the change in the log Ka for K^+ caused by the addition of Na^+. This is summarized in Table VI. We now have a situation where the competitive binding by Na^+ does not affect the signal but decreases the log Ka values for K^+ binding. Decreases in the log Ka implies that the ionophore binds K^+ weakly but more reversibly. Thus, MCC222 performs better as a sensor in the presence of Na^+ than it does in its absence. However, it will be less sensitive.

In medicinal chemistry, Na^+ will be described as an antagonist since it produces no effect of its own on the fluorescence and yet competes with K^+ which is an agonist. K^+ does affect the fluorescence when bound to the sensor (see Appendix II). The apparent Ka values for K^+ binding in Table VI were used to calculate the selectivity of K^+ over Na^+ for binding MCC222. Using a linear regression analysis (see Appendix II), this value was determined to be 15. The calculated Ka for Na^+ is 35 M^{-1} which corresponds to log Ka of 1.54. Earlier it was indicated that the selectivity of K^+ over Na^+ of the order of 500 was necessary for the fluorescent chemosensors to be effective in continuous monitoring devices. However, this constraint was based on the assumption that Na^+ will behave like an agonist. Since it is an antagonist it does not interfere with the transduction and therefore, a selectivity of 15 is sufficient for practical applications. Further, fluctuations in the concentration of Na^+ during open-heart surgery will not affect the apparent Ka value for K^+ as seen in Table VI. If necessary, correction can be made for such fluctuations.

The competitive binding of Na^+ to MCC222 was confirmed by FABMS method. The addition of NaCl and KCl to MCC222 in the glycerol matrix showed a predominant peak at 545 corresponding to $(M+K^+)$ and minor peak at 529 $(M+Na^+)$.

We selected MCC222 for developing the fiber optics-based sensor for monitoring K^+ in the extracorporeal blood during open-heart surgery. The technology has been licensed to a diagnostic company and commercialization is expected within the next couple of years.

Microfluorometric Assay for K^+. We have already shown that the log Ka value for K^+ in methanol for MCC222 is two orders of magnitude larger than the value in water. This implies that the ionophore is more strongly bound to K^+ in methanol than in water. Therefore, it may behave as a reagent in methanol. As a fluorescent reagent, MCC222 can be used in a continuous flow microfluorometric assay (21). This technique is so sensitive that only nanoliter quantities of samples are needed for analysis. The feasibility of using MCC222 as a reagent was demonstrated as follows: the flow assay solution was prepared by dissolving MCC222 (7.6 μM) and NaCl (1.4 mM) in 100% methanol. Two sets of analytical standards were prepared in water. These standards represent the blood serum samples. In both sets, the level of KCl varied from 1.0-6.0 mM. The first set also contained NaCl at a constant level of 50 mM. In the second set, the level of NaCl was maintained at 150 mM. The level concentration of Na^+ in human blood is approximately 135 mM. The 50 mM and 150 mM levels of NaCl therefore represent extreme limits in NaCl content. Our purpose here is to demonstrate that changes in the level of Na^+ in blood samples will not affect our assay, and that the reagent (MCC222) will respond only to concentration changes in K^+. Because of the strong binding of the reagent to K^+ in methanol, the standard samples had to be diluted 2000:1 to bring the concentration of K^+ within the "dynamic range". The results are summarized in Table VII. Whether the level concentration of Na^+ is 50 mM or 150 mM, the fluorescence intensity depended only on the concentration of K^+. The coefficient of variation was of the order of 0.96.

Table VII also shows the ratios of the reagent ionophore concentrations to K^+ concentrations for the samples were >1. This is a proof that the reagent is now consumed in stoichiometric quantities. Further, in Table VII, the ratio concentrations of Na^+ to K^+ for the samples varies from 3000 to 500. This confirms the fact that in methanol, MCC222 is more selective and sensitive towards K^+ than in water.

Table V. Properties of 4-Methylcoumaro [222] Cryptand

EXCITATION PEAK, NM	EMISSION PEAK, NM	LOG $(KA)_{K+}$ (H_2O)	LOG $(KA)_{K+}$* $(MeOH)$	QUANTUM YIELD[A] (RELATIVE)
340	420	2.72	4.92	0.045

LOG $(KA)_{K+}$ = LOG ASSOCIATION CONSTANT

* = THIS MEASUREMENT WAS MADE IN THE PRESENCE OF 1.4 MM Na^+ IN 100% MeOH

[A] = RELATIVE TO QUININE SULFATE (0.1 ABSORBANCE IN 1 N SULFURIC ACID)

Table VI. Effect of Na^+ on Log $(KA)_{K+}$ of 4-Methylcoumaro [222] Cryptand

	Na^+ CONCENTRATION, MM				
	0	10	50	100	140
LOG $(KA)_{K+}$	2.72	2.64	2.34	2.03	1.97

Table VII. Microfluorometric Assay for K^+

K^+		Ratio	Fluorescence Intensity (arbitrary units)	
Assay Solution (mM)	After Dilution[1] (mM)	Ionophore /K^+	Assay Solution (I)[2] 50 mM Na^+	Assay Solution (I$_S$)[3] 150 mM Na^+
1.0	0.0005	15.2	18.1 ± 0.25^4	17.6 ± 0.46
2.0	0.0010	7.6	35.6 ± 0.32	35.6 ± 0.24
3.0	0.0015	5.1	52.5 ± 0.44	52.4 ± 0.48
4.0	0.0020	3.8	68.9 ± 0.75	68.4 ± 0.53
5.0	0.0025	3.0	84.7 ± 0.42	83.7 ± 0.36
6.0	0.0030	2.5	100.3 ± 0.42	99.6 ± 0.47^5

1. Diluted 1:2000 in a flowing assay solution of 4-methylcoumaro[222] cryptand (7.6 µM) and NaCl (1.4 mM) in methanol
2. Total Na^+ 1.425 M
3. Total Na^+ 1.475 M
4. Na^+ upper limit ≈ 3000-fold excess
5. Na^+ lower limit ≈ 500-fold excess

K⁺ Transport Across Biological Membranes. MCC222 is also an excellent research tool for studying the rate of transport of K⁺ across biological membranes. This work was carried out in collaboration with Prof. Ira Kurtz of the UCLA College of Medicine and published earlier (22). This work will not be discussed here.

Mechanism of Fluorescence Enhancement in K⁺-Bound 4-Methylcourmaro[222]-cryptand. Ion-binding quenches the fluorescence in coumaro-crown ethers while it enhances the fluorescence in coumaro-cryptands. Between the two effects, fluorescent enhancement is more dramatic than quenching. Even though higher selectivities of the cryptands may account in part for the higher sensitivities, the reason why these effects are opposite for the two systems is intriguing.

We have already indicated that the two nitrogens in the cryptand series may affect the fluorescence in ways different from the crown ether series. The crown ethers are not sensitive to changes in pH. However, changing the pH from 7.0 to 7.4 decreases the fluorescence emission of MCC222 by 16% in the absence of K⁺. In its presence, the fluorescence was increased by 12%. This lends further support that the nitrogens indeed play a critical role in fluorescence enhancement. Deprotonation of the nitrogens (due to pH increase) would expose the lone electron pairs of these nitrogens. Such deprotonation suppresses the fluorescence. Presumably, deprotonation also promotes K⁺ binding; the lone pairs of electrons of the nitrogens are no longer exposed and an enhancement in fluorescence results.

The extinction coefficient (ϵ) of all the six coumaro-ionophores are of the same order (10^{-4}) and are therefore equally efficient in absorbing light. However, the coumaro-cryptands are an order of magnitude less efficient in their fluorescence emission than the coumaro-crown ethers. The quantum yield (relative to quinine sulfate) for MCC222 in water is 0.045. (See Table VI.) The optimum concentration of the coumaro-crown ethers in 50:50 ethanol-water is 10^{-5} M while it is 10^{-4} M for coumaro-cryptands.

Our explanation as to why the fluorescent enhancement of MCC222 is dramatically increased on addition of K⁺ is summarized in Figure 7. When this cryptand is not bound to K⁺, the singlet dipolar excited state (17,18) is susceptible to the transfer of a single electron from the nitrogen proximal to the positively charged oxygen at position 7. This photo-induced electron transfer (PET) converts the dipolar singlet excited state to a radical ion pair, in which the radical cation resides on the nitrogen while the anion radical is delocalized at the α,β-unsaturated ester functionality. These two radicals are no longer conjugated, therefore undergo a radiationless decay. Thus, intramolecular PET causes the fluorescence to be suppressed in the free ionophore.

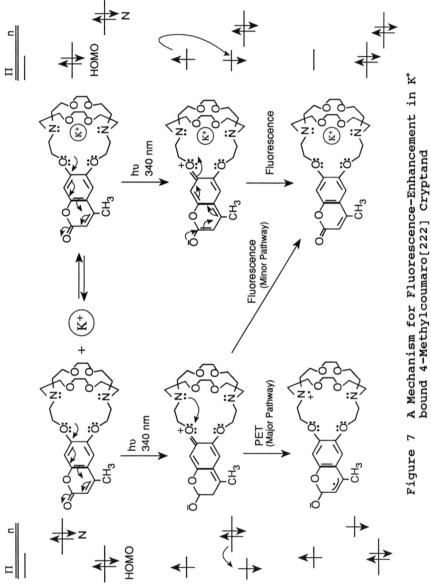

Figure 7   A Mechanism for Fluorescence-Enhancement in K⁺ bound 4-Methylcoumaro[222] Cryptand

When the cryptand is bound to K^+, the nitrogen lone pairs are stabilized and their ionization potentials are increased. PET is no longer favored and fluorescence emission is "turned-on" since the pathway suppressing fluorescence is "turned-off". Similar chelation-enhanced fluorescence has been reported by Czarnik (23) and de Silva (24). However, this triggering of fluorescence does not happen for Na^+ since MCC222 is not selective for this ion. The Na^+ binding does not stabilize the nitrogen lone pairs sufficiently to prevent PET. Fluorescence enhancement is, therefore, not observed for this ion.

In the case of coumaro-crown ethers, there are no nitrogens available for PET and, therefore, the fluorescence is not suppressed in the free ionophore. However, metal ion binding quenches the fluorescence through distortions caused by the repulsive interaction between the positively charged oxygen at position 7 and the metal ion. Of course, this effect is present in the cryptands also, but is overwhelmed by the PET in the unbound state.

Conclusion

We have invented a fluorescent chemosensor for *in vivo* monitoring of K^+ in blood. By changing the solvent, the same chemosensor can be used as a fluorescent reagent in automated batch analysis of K^+. The chemosensor also lends itself as a research tool in studying K^+ transport across biological and synthetic membranes. But most importantly, we have demonstrated that this information pertaining to ion-binding in receptors can be communicated instantaneously in a cost-efficient way. This understanding opens the door for further adventure into areas of molecular electronics, molecular recognition and information storage and retrieval.

Acknowledgements

Dr. Ira Kurtz of the UCLA College of Medicine was an active collaborator in this project and provided many of the spectral data. His contribution is gratefully acknowledged. We thank Ray Brambilla who performed the NMR work. We appreciate the work of David Hindenlang and Don Sedgwick who provided FAB mass spectral data. Ken Legg supported our work and recognized the importance of the fluorescence approach to K^+ sensing. George S. Hammond shepherded the project and was a source of inspiration. The contributions of Ken and George are greatly appreciated. Marylou Grumka is acknowledged for typing the manuscript.

Appendix I. Synthesis of Coumarin-Fused Ionophores

Appendix II

The referee who reviewed this paper brought to our attention terms "agonist" and antagonist" used by scientists in the field of medicine.
The referee also helped us to calculate the selectivity of K^+ over Na^+ by MCC222 using the apparent log Ka values for K^+ determined in the presence of Na^+ and reported in Table VI.
Because

$$^1/\text{Apparent Ka for } K^+ = \frac{[K^+][\text{Free Ligand+Na}^+.\text{Ligand}]}{[K^+.\text{Ligand}]}$$

and since $[Na^+.\text{Ligand}] = [\text{Free Ligand}] [Na^+] (Ka)_{Na^+}$,

$$^1/\text{Apparent Ka for } K^+ = \frac{1 + [Na^+](Ka)_{Na^+}}{(Ka)_{K^+}} \text{ where } (Ka)_{K^+} \text{ is the}$$

true Ka for K^+.
Thus, a plot of $^1/$Apparent Ka for K^+ vs $[Na^+]$ should give a straight line and the slope of this line will give selectivity Na^+/K^+. This linear regression analysis using data from Table VI gave a slope of 0.067. The reciprocal of this value, which is 15, corresponds to the selectivity of K^+ over Na^+ for MCC222. Using this value, $(Ka)_{Na^+}$ for MCC222 was calculated to be 35 M^{-1} (or, the log $(Ka)_{Na^+} = 1.54$).
We thank the referee for this contribution.

Literature Cited

1. First reported by Stinson, S. Chem. Eng. News **1987,** 65 (45), 26. Patents covering this work: Masilamani, D., Lucas, M. E., Hammond, G. S.; Fluorogenic and Chromogenic Three-Dimensional Ionophores as Selective Reagents for Detecting Ions in Biological Fluids, U. S. Patent 5,162,525, 1992; Chem. Abstr. **1990,** 112, 51780 and Ion Selective Fluorogenic Reagents, U. S. Patent 5,136,033, 1992.
2. Smith, R. V.; Nessen, M. A. J. Pharm. Sci. **1971,** 60, 907-908.
3. Ekins, R. P. Clin. Biochem. Rev. **1987,** 8, 12-23.
4. Freundlich, N. J. Ind. Chem. News **1986,** 7, 1.
5. Miller, J. N. Chemistry & Industry **1984,** 22-29.
6. Narayanaswamy, R. Anal. Proc. **1985,** 22, 204-206.
7. Thompson, M.; Krull, U. J. Anal. Chem. **1991,** 63, 393A-405A.
8. Grell, E.; Lewitzki, E.; Hoa, D. H. M.; Gerhard, A.; Rup, H.; Krause, G.; Mager, G. Ion Transport Through Membranes; Academic Press: New York, 1987; pp 41-60.
9. Dietrich, B. J. Chem. Ed. **1985,** 62, 954-964.
10. Kolthoff, I. M. Anal. Chem. **1979,** 51, 1R-22R.
11. Izatt, R. M.; Bradshaw, J. S.; Nielsen, S. A.; Lamb, J. D.; Christensen, J. J. Chem. Rev. **1985,** 85, 271-339.
12. Gokel, G. W.; Goli, D. M.; Minganti, C.; Echegoyen, L. J. Am. Chem. Soc. **1983,** 105, 6786-6788.

13. Lehn, J. M. Pure Appl. Chem. **1977, 49**, 857-870.

14. Kumar, A.; Chapoteau, E.; Czech, B. P.; Gebauer, C. R.; Chimenti, M. Z.; Raimondo, O. Clin. Chem. **1988, 34**, 1709-1712.

15. Pedersen, C. J. J. Am. Chem. Soc. **1967, 89**, 7017-7035.

16. Kyba, E. P.; Helgeson, R. C.; Madan, K.; Gokel, G. W.; Tarnowski, T. L.; Moore, S. S.; Cram, D. J. ibid **1977, 99**, 2564-2571.

17. Jones, II, G.; Griffin, S. F.; Choi, C-y.; Bergmark, W. R. J. Org. Chem. **1984, 49**, 2705-2708.

18. Abu-Eittah, R. H.; El-Tawil, A. H. Can. J. Chem. **1985, 63**, 1173-1179.

19. Rettig, W.; Klock, A. ibid **1985, 63**, 1649-1653; Seypinski, S.; Drake, J. M. J. Phys. Chem. **1985, 89**, 2432-2435.

20. The Merck Index; Stecher, P. G. (Ed); Merck & Co.: Rahway, NJ, 1968; p 290.

21. Good, D. W.; Vurek, G. G. Anal. Biochem. **1983, 130**, 199-202.

22. Golchini, K.; Mackovic-Basic, M.; Gharib, S. A.; Masilamani, D.; Lucas, M. E.; Kurtz, I. Am. J. Physiol. (Renal Fluid Electrolyte Physiol. 27) **1990, 258**, F438-F443.

23. Huston, M. E.; Haider, K. W.; Czarnik, A. W. J. Am. Chem. Soc. **1988, 110**, 4460-4462.

24. de Silva, A. P.; Gunaratne, N.; Sandanayake, K. R. A. S. Tetrahedron Lett. **1990, 31**, 5193-5196; de Silva, A. P.; Sandanayake, K. R. A. S. ibid **1991, 32**, 421-424.

RECEIVED July 9, 1993

Chapter 12

Fluorescent Probes in Studies of Proteases

Grant A. Krafft and Gary T. Wang

Structural Biology, Drug Design and Delivery, Pharmaceutical Product Division, Abbott Laboratories, Abbott Park, IL 60064-3500

Proteases are involved in a wide spectrum of biological functions, including regulatory, catabolic, metabolic and pathologic processes. We have designed and synthesized internally quenched fluorogenic protease substrate probes that enable enzymology of purified, characterized proteases such as HIV protease and renin, which are targets for therapeutic intervention in AIDS and hypertension, respectively. We have also developed fluorogenic substrate probes that permit the search for unknown or uncharacterized proteases with putative cleavage functions that may be associated with particular disease processes, such as ß/A4 amyloid formation in Alzheimer's disease. In this chapter, applications of these substrate probes in studies pertaining to HIV protease, renin, and putative Alzheimer's disease associated proteases are described.

Protease enzymes are essential participants in many biological processes, ranging from relatively non-specific degradation of proteins and protein fragments, as in the ATP-dependent ubiquitin proteolysis pathway (1), to highly specific processes such as the generation of important biological mediators such as angiotensin (2) and endothelin (3). Proteases also play a significant role in disease processes mediated by pathogens, as exemplified by the degradation of hemoglobin by a malarial protease (4), or the processing of virally encoded polyproteins by HIV protease to produce mature and infectious HIV particles (5). The enzymologic characterization of many proteases, such as thrombin, plasmin, and tissue plasminogen activator has been accomplished successfully and with relative ease using small chromogenic or fluorogenic di- or tripeptide substrates (6). While these substrates are clearly valuable in studies of many proteases, they are inadequate for studies of proteases which require a higher level of molecular recognition and interaction with amino acid residues on both sides of the substrate cleavage site. For proteases with these extended molecular recognition requirements, internally quenched fluorogenic substrates containing relatively long peptide sequences have been developed in our laboratories (7-10), enabling enzymologic characterization and protease inhibitor screening, but also facilitating the more interesting and important identification of protease enzymes involved in newly discovered biological proteolytic events.

0097-6156/93/0538-0183$06.00/0

Design and Synthesis of Fluorogenic Protease Substrates

An ideal protease assay should be continuous, fast and adaptable to microtiter plate format for high through-put data generation. Thus, several challenges confront the designer of substrate probes for proteases. The first challenge is to achieve continuous signal generation as a direct result of the proteolysis event. The second challenge is the incorporation of an appropriate peptide sequence that will be recognized and efficiently cleaved by the target protease. A third challenge involves optimization of substrate attributes that enhance their practical utility, such as aqueous solubility and desirable spectral parameters. Finally, substrate synthesis must be fast and efficient in order to capitalize on substrate sequence variation in studies of protease molecular recognition preference.

Continuous Signal Generation. Signal generation in short, di- and tripeptide chromogenic or fluorogenic substrates that include only P side amino acid residues relies on electronic changes in a chromogenic or fluorogenic moiety when proteolytic deacylation occurs. A prototypic chromogenic substrate for thrombin (11) and a prototypic fluorogenic substrate for urokinase (12) are shown below (Figure 1.). Proteolysis generates p-nitroaniline and 7-amino-4-methylcoumarin, coincident with measurable increases in optical density and fluorescence intensity, respectively.

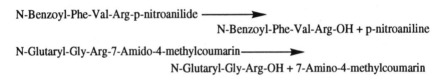

N-Benzoyl-Phe-Val-Arg-p-nitroanilide ⟶

    N-Benzoyl-Phe-Val-Arg-OH + p-nitroaniline

N-Glutaryl-Gly-Arg-7-Amido-4-methylcoumarin ⟶

    N-Glutaryl-Gly-Arg-OH + 7-Amino-4-methylcoumarin

Figure 1. Signal generation in peptidic p-nitroanilides (chromogenic) or 7-amino-4-methylcoumarins (fluorogenic). These substrates contain amino acid residues only on the P side. with none on the P' side.

Proteolysis that occurs between two amino acid residues of a generic peptide sequence does not offer an analogous opportunity for signal generation, since no easily measurable spectral change occurs for this process. Thus, our solution to the signal generation challenge was based on the concept of resonance energy transfer that can occur between a fluorophore and a suitably situated acceptor chromophore as illustrated in Figure 2. This concept had been employed almost two decades ago by Latt and coworkers (13), who prepared carboxypeptidase A substrates of the type Dns-(Gly)n-Trp, n=1-3, in which the dansyl group served as an energy acceptor that quenched the tryptophan fluorescence. The residual fluorescence of the substrate containing a single Gly residue was 1%, but increased to 25% with three Gly residues. Thus, the dynamic range for signal generation upon complete liberation of the tryptophan fluorophore for the longest substrate was only 3:1. This suggested that a much more effective fluorophore-quencher pair would be required for efficient quenching in substrates consisting of five or more amino acid residues.

In order to discover a fluorophore-quencher pair with optimized quenching, several conjugates of typical fluorophores and acceptor chromophores were prepared (7). The most efficient quenching was observed for the conjugate containing the fluorophore Edans (5-(2'-aminoethylamino)naphthalene sulfonate) and the acceptor chromophore Dabcyl (4-(4'-dimethylaminobenzeneazo)benzoyl), for which the residual fluorescence was much less than 0.5%. This pair was selected because of the excellent spectral overlap of the Edans emission and Dabcyl absorbance, as shown in Figure 2. The fluorescence quenching of this fluorophore-quencher pair is further enhanced by

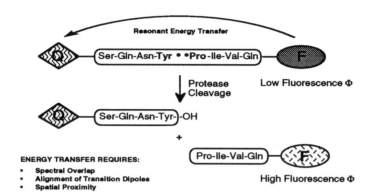

Figure 2. Internally quenched fluorogenic protease substrate enable signal generation by liberation of the fluorophore from close proximity with the quenching chromophore. Efficient quenching is achieved due to the excellent spectral overlap of the Edans fluorophore and the Dabcyl chromophore.

the relatively long fluorescence life time (ca. 15 ns). It was expected that this optimized overlap would translate into efficient energy transfer, even when the separation of quencher and fluorophore was extended to six, eight or even ten amino acid residues.

Substrate Peptide Sequence Selection. For many proteases of interest, some information regarding substrate cleavage specificity is available, usually from sequence identification of peptide or protein fragments believed to be the primary products of the protease hydrolysis reaction. The selection of the specific peptide sequence that is incorporated into a particular protease substrate is usually the shortest sequence anticipated to enable adequate recognition by the protease, since quenching efficiency, and ultimate assay sensitivity will be maximized when the minimal peptide length is used. Spacing residues can be inserted between the minimum recognition sequence and the chromophores to minimize repulsive steric interactions at the enzyme active site, in instances where this may be a problem. Consideration is also given to incorporation of additional polar or charged residues that might exist near the putative scissile bond, to impart additional water solubility (*vide infra*).

Fluorogenic Substrate Synthesis. The first internally quenched fluorogenic substrates that we prepared several years ago were obtained by simple carbodiimide or active ester-mediated amide formation between the amino and carboxyl residues of an octapeptide (Ser-Gln-Asn-Tyr-Pro-Ile-Val-Gln), and Edans and Dabcyl, respectively (7, 8). This method was reasonably efficient, requiring initial solid phase synthesis and purification of the desired peptide, two solution phase coupling steps and HPLC purification of the final substrate. However, if the peptide to be functionalized in this manner contained side chains bearing nucleophilic groups or carboxyl groups, additional protection and deprotection steps would be required. To avoid this additional chemistry, and to enhance the overall efficiency of the synthesis, we developed a fully automated synthetic protocol utilizing Fmoc chemistry on a standard solid phase peptide synthesizer (9). This method relies on the two derivatized Fmoc-amino acids shown below (Figure 3), which incorporate the Edans and Dabcyl groups onto the side chains of glutamic acid and lysine, respectively. This automated method allows the synthesis of substrates containing virtually any peptide sequence of virtually any length that can be reasonably constructed by solid phase synthetic methods. Moreover, this method permits incorporation of the fluorophore and quenching

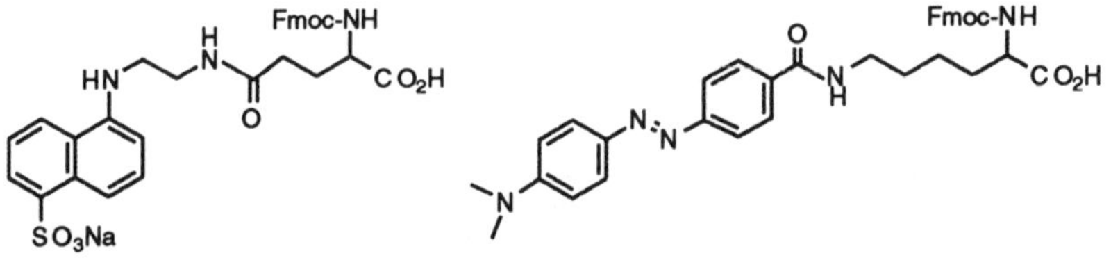

Figure 3. Structure of N_α-Fmoc-Glutamic acid-γ-Edans (Glu-Edans) and N_α-Fmoc-N_ε-Dabcyl lysine (Lys-Dabcyl).

chromophore at any position of a peptide, not only at the termini. This permits incorporation of water soluble, charged amino acids at the ends of the substrate and external to the chromophores, a configuration which does not increase the distance between the fluorophore and quencher, and which minimizes the possibility that the charged residues would adversely affect substrate binding.

Fluorogenic Substrate Probes for Studies of HIV Protease

HIV protease (HIV PR) is a C-2 symmetric, dimeric protease enzyme consisting of 11 kDa monomers encoded by HIV as part of the gag-pol fusion polyprotein (14-16). It is responsible for processing the gag and gag-pol polyprotein precursors to produce mature, infectious HIV particles. Over the past five years, HIV PR has been the focus of intense investigation, since it provides a promising target for therapeutic intervention in AIDS (17, 18). X-ray crystallographic studies have determined the three-dimensional structure of HIV PR (19), and the structures of several HIV PR-inhibitor complexes (20-21). Many studies directed towards the discovery of HIV PR inhibitors have resulted in the development of extremely potent protease inhibitors, some of which exhibit promising anti-viral activity in cell culture studies (22-24).

Peptide Sequence Selection for HIV PR Substrates. At least eight distinct proteolytic processing sites are cleaved within the gag and gag-pol polyproteins by HIV PR (25-28) and three autodegradation cleavages within the HIV PR have been identified (29). These are delineated in Figure 4 with residues at the scissile bond shown in boldface. These sequence provide the initial information pertaining to substrate peptide sequence that is needed to design a substrate. Two motifs in and around the scissile bond are evident: aromatic-proline sites or hydrophobic-hydrophobic sites. For the initial substrate that was prepared in our laboratories, we selected the p17*p24 site, since HPLC studies with simple peptide has shown (30) that this peptide was cleaved efficiently by HIV PR. Subsequently, we have prepared substrates corresponding to a number of additional sites.

gag 132/133 (p17*p24)	Val-Ser-Gln-Asn-**Tyr-Pro**-Ile-Val-Gln-Asn
gag 363/364 (p24*X)	Lys-Ala-Arg-Val-**Leu-Ala**-Glu-Ala-Met-Ser
gag 377/378 (X*p7)	Thr-Ala-Thr-Ile-**Met-Met**-Gln-Arg-Gly-Asn
gag 448/449 (p7*p6)	Arg-Pro-Gly-Asn-**Phe-Leu**-Gln-Ser-Arg-Pro
pol 68/69 (*PR)	Val-Ser-Phe-Asn-**Phe-Pro**-Gln-Ile-Thr-Leu
pol 167/168 (PR*RT)	Cys-Thr-Leu-Asn-**Phe-Pro**-Ile-Ser-Pro-Ile
pol 607/608 (RT*RNase H)	Gly-Ala-Glu-Thr-**Phe-Tyr**-Val-Asp-Gly-Ala
pol 727-728 (RT66*IN)	Ile-Arg-Lys-Ile-**Leu-Phe**-Leu-Asp-Gly-Ile

Figure 4. Sequences of HIV cleavage sites. Amino acid residues were numbered according to the sequence of HIV *gag-pol* polyprotein, with residues at the scissile bond shown in boldface.

Hydrolysis of Fluorogenic HIV PR Substrates by HIV PR. When the fluorogenic substrate Dabcyl-Gaba-Ser-Gln-Asn-Tyr-Pro-Ile-Val-Gln-Edans, **1**, was incubated with purified, recombinant HIV PR, a rapid, linear increase in fluorescence intensity is observed. A typical hydrolysis reaction is presented in Figure 5. Complete hydrolysis of this substrate resulted in a 40 fold signal enhancement, indicating a fluorescence quenching efficiency of 97.5%. This result was quite gratifying, since this substrate represented the longest internally quenched peptide substrate that had been prepared to that point, with thirty-two atoms separating the fluorophore and acceptor chromophore. This quenching efficiency also suggested that the actual time-averaged distance between fluorophore and acceptor chromophore (over the life time of the EDANS excited state) was significantly less than the extended peptide conformation, based on calculation of quenching efficiency using the Forster equation (8).

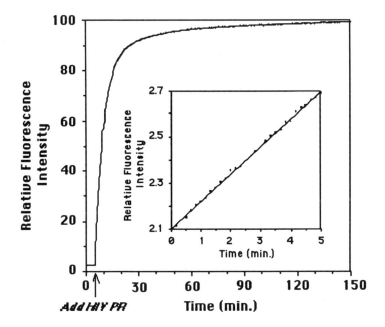

Figure 5. Hydrolysis of fluorogenic substrate 1 by HIV protease at 37 °C as monitored by steady state fluorescence spectroscopy ($\lambda_{ex}+340$, $\lambda_{em}=490$). The reaction was carried out with 10 μM substrate at pH 4.7 in a buffer solution containing 0.1 M NaOAc, 1.0 M NaCl, 1 mM EDTA, 1 mM dithiothreitol, 10% DMSO and 1 mg/mL bovine serum albumin. The arrow indicates the point of addition of HIV protease to a final concentration of 35 nM. Product analysis was carried out by HPLC, mass spectrum and fluorescence lifetime. Inset: The initial phase of the hydrolysis reaction used for rate determinations.

A variety of substrates for HIV protease, encompassing different scissile bonds within the HIV polyproteins, have been prepared, and their kinetic parameters have been measured. Selected data are presented in Table I below. All of the substrates that we have prepared are cleaved with good efficiency. The sequence encompassing the N-terminus of HIV PR itself (VSFNFPQIT) has the lowest K_m (1.7 μM), while the sequence encompassing the p17*p24 cleavage site (SQNYPIVQ) is cleaved the fastest

(9.5 sec^{-1}). The most efficiently cleaved HIV protease substrate reported in the literature to date consists of the sequence Lys-Ala-Arg-Val-Leu*-p-nitroPhe-Glu-Ala-norLeu-NH$_2$, with a k_{cat}=32 sec^{-1} (31).

The fluorogenic substrates enabled competition experiments with unfunctionalized peptide substrates and with inhibitors. Table II presents data for competitive inhibition of HIV protease by unfunctionalized peptides. In these experiments, equimolar concentration of unfunctionalized peptide and substrate were incubated with HIV PR under reaction conditions described in Figure 5. Three of the sequences exhibited significant inhibitory capacity, and paralleled the K_m data presented in Table I. (the K_m and V_{max} for the TATIMMQRGE substrate were compared to the norleucine substrate presented in Table I). A surprising result that emerged from these experiments was the fact that the equimolar concentrations of the unfunctionalized peptide SQNYPIVQ and TATIMMQRGE inhibited their counterpart fluorogenic substrates only ca. 13-15%, instead of the 50% that would be expected for ligand with identical binding capability. This suggested that the fluorophore and quencher attached at either end of the peptide substrate may be contributing to enhanced enzyme-substrate binding interactions. Further evidence for enhanced binding contribution by the fluorophore and quencher group is evident from comparison of the 70 μM K_m value of the SQNYPIVQ fluorogenic substrate, versus the higher values of 2.3 mM reported by Billich and Winkler for unfunctionalized peptide (32). Another potential contributing factor for these observations may be the higher solubility of the fluorogenic substrates compared to the simple peptide substrates which may form soluble aggregates that are not readily available for proteolysis.

Table I. Kinetic Parameters for Fluorogenic HIV Protease Substrates [a]

Substrate Sequence	K_m (μM)	V_{max} (μM/min)	k_{cat} (sec^{-1})	k_{cat}/K_m
Dab-Gaba-SQNYPIVQ-Ed	70	0.18	9.5	136,000
R-J-TATInLnLQRG-B-R	7.0	0.17	---	---
Dab-HVSFNFPQITH-Ed	1.7	0.009	1.5	884,000
R-J-VSFNFPQIT-B-R	9.8	0.008	1.5	153,000
Ac-R-J-SQNYPIVQ-B-R-NH$_2$	44	0.024	4.7	106,000
RR-J-SQNYPIVQ-B-RR	43	0.038	2.7	63,500

[a]In substrate sequence, "J" denotes Glu-Edans and "B" denotes Lys-Dabcyl. Initial rates were determined as described in Figure 5. At least six concentrations were used for K_m and V_{max} determinations, and all hydrolysis reactions were run in triplicate. All substrates exhibited typical Michaelis-Menten kinetics. At substrate concentrations higher than 20 μM, the inner filter effect became significant, requiring a correction for light absorbed by substrate (8).

Fluorogenic Substrate Assay for Inhibitors of HIV PR. The development of fluorogenic HIV PR substrates was motivated by the need to evaluate the inhibitory potency of potential drug candidates for anti-HIV therapy. A 96-well microtiter plate adaptation of the assay published in 1990 (8) has facilitated screening and evaluation of large number of inhibitor molecules, including more than 1000 synthetic analogs. In cases where the inhibitor concentration needed to inhibit 50% of the enzyme (IC$_{50}$) approaches the concentration of the enzyme used in the assay, the values are simply

Table II. Competition Inhibition of HIV Protease Cleavage of Fluorogenic
Substrates by HIV peptides [a]

Peptide	Dab-SQNYPIVQ-Ed	Dab-TATIMMQRGE-Ed
SQNYPIVQ	3.5%	8.7%
TATIMMQRGE	24.6%	12.7%
RVSFNFPQITR	29.6%	22.6%
ATLNFPISQE	4.8%	--
IRQANFLGL	8.0%	--
SVVYPVVQ	1.3%	0%
TFQAYPLRGA	1.3%	--

[a]Competition experiments conducted as described in Figure 5, with peptide concentration also at 10 µM. Data are expressed as percent of initial velocity for the cleavage reaction of fluorogenic substrates in the absence of unfunctionalized peptide.

stated as <1 nM. More accurate values and actual K_i's can be determined by kinetic analysis appropriate for tight binding ligands.

Fluorogenic Substrates in Studies of Renin

Like HIV protease, renin is a member of the aspartic protease family. It plays an important role in blood pressure regulation and electrolyte homeostasis through the renin-angiotensin system (33). Upon release into the bloodstream, renin specifically hydrolyzes the glycoprotein angiotensinogen to form angiotensin I, which in turn is hydrolyzed by angiotensin converting enzyme to generate angiotensin II. This octapeptide is a potent vasoconstrictor and a promoter of aldosterone release and sodium retention (34). High renin activity levels also have been linked directly with high risk of heart disease (35). Thus, renin represents an attractive target for therapeutic intervention of hypertension.

Traditionally, renin activity has been determined using a time point assay by radioimmunoassay of angiotensin I produced from a radiolabeled tetradecapeptide substrate (36, 37). A kinetic profile of renin activity can be obtained by making sequential RIA measurements, though this is impractical for assaying large numbers of renin reactions, and impedes detailed kinetic analysis of renin. Several other types of assays for renin activity also have been published. A fluorescence assay based on a C-terminal amide of ß-naphthylamine or 7-amino-4-methylcoumarin has been described (38, 39), in which renin carries out the initial endoproteolytic event to generate a carboxy terminal fragment, which in turn is degraded by the action of an aminopeptidase enzyme to free the fluorescent label. This coupled fluorescence assay is potentially simpler and easier to use than the RIA, but it is not a continuous assay and it requires multiple time point measurements to assess enzyme kinetic parameters. An HPLC-based assay employing a synthetic tetradecapeptide substrate suffers from the same drawback (40). Thus, substrates based on internal fluorescence quenching, analogous to the fluorogenic HIV protease substrates, were an attractive alternative.

The proteolytic action of human renin is remarkably specific, recognizing a peptide sequence Ile-His-Pro-Phe-His-Leu-Val-Ile-His-Thr which contains the P_6-P_3' residues spanning the normal Leu-Val cleavage site of human angiotensinogen (39). We prepared two different fluorogenic substrates based on this recognition sequence, as shown on the following page.

Dabcyl-gaba-Ile-His-Pro-Phe-His-Leu-Val-Ile-His-Thr-Edans

2

Ac-Arg-Glu(Edans)-Ile-His-Pro-Phe-His-Leu-Val-Ile-His-Thr-Lys(Dabcyl)-Arg-NH$_2$

3

The first substrate **2** was synthesized via solution chemistry (7, 10), and the second substrate **3** was synthesized via the automated solid phase procedure, incorporating two arginine residues at the periphery to enhance the substrate solubility (9). Detailed kinetic characterization of substrate **2** has been carried out (10, 41-42).

Substrate Hydrolysis by Renin. When substrate **2** was incubated with recombinant human renin, two peptide fragments were generated. HPLC analysis showed one fluorescent fragment with no absorption at 490 nm, and one non-fluorescent fragment with maximal absorption at 490 nm (41). Mass spectrometry analysis and amino acid sequencing indicated that the fluorescent fragment was VIHT-Edans and the non-fluorescent fragment was Dabcyl-gaba-IHPFHL, thus confirming the specific cleavage of the fluorogenic substrate at the predicted Leu-Val site.

Analysis of the hydrolysis kinetics of substrate **2** by recombinant human renin revealed competitive substrate inhibition (42). This model postulates that the first molecule of substrate binds the enzyme in a catalytically productive fashion forming the ES complex. At higher substrate concentrations, a second molecule of substrate binds to the active ES complex to form inactive ES$_2$ complex, resulting in rate reduction. Kinetic analysis using this model gave a K_m of 3.6 µM and a K_i for inhibition by a second substrate molecule of 422 µM. The pH-activity profile of renin-catalyzed hydrolysis of the fluorogenic substrate indicated a pH optimum at about pH 8.0, a shift of 2.6 pH units compared with the pH optimum for renin cleavage of angiotensinogen (pH 5.4) (33).

The sensitivity of the assay using this fluorogenic substrate was evaluated by conducting a series of enzyme dilution experiments (10). A detection level as low as 30 ng/mL after incubation of only 3-5 minutes was measured. With extended incubation time (2-3 h), it was estimated that the assay could detect renin at 0.5 ng/mL.

The renin substrate has been adapted to a time point format on 96-well microtiter plate fluorescence reader for high-throughput data generation. This format has been used to examine the effect of a series of reagents (DMSO, high salt, DTT, etc.) on the stability and activity of recombinant human renin.

Fluorogenic Probes for Studies of Alzheimer's Disease-Related Proteases

The fluorogenic substrates for HIV protease and renin represent probe molecules for studies of known enzymes. These substrates have proven valuable in a number of applications, including enzyme purification, kinetic studies, inhibitor characterization and inhibitor screening. Another potential application for this type of probes is to search for unknown or uncharacterized proteases with putative cleavage functions associated with particular physiologic or disease processes. Fluorogenic substrates based on established or speculated cleavage site sequences can be prepared and utilized as tools to characterize and purify the relevant protease activity from appropriate biological samples (tissue homogenate, cell lysate, etc.). An illustrative example of this strategy is our recent effort to identify and purify proteases involved in processing of amyloid precursor protein (APP) and generation of the amyloid peptide found in Alzheimer's disease.

In 1987 it was established that the 39-43 residue Aβ peptide was encoded by the gene for a larger protein, the amyloid precursor protein (APP) (43-45). Alternate mRNA splicing of this gene produces a number of APP isoforms (APP770, APP751, APP714, APP695), all of which are glycosylated, integral membrane proteins (46). A secretory processing pathway for APP transfected into HEK 293 was shown to involve a cleavage between Lys_{612} and Leu_{613}, a site within the Aβ peptide, that prevented the possibility of Aβ formation (47). Since the termini of Aβ had been established (48), it was straightforward to generate internally quenched fluorogenic substrate probes corresponding to the cleavage sites at the Aβ termini, to identify the proteases that were involved in generating Aβ from APP by a non-secretory route. The following section describes preliminary experiments employing these substrates to elucidate the amyloidogenic proteases from brain tissue samples.

Substrate Design and Synthesis.   The octapeptide sequences that encompass the N-terminal amyloidogenic cleavage site and the secretory cleavage site of APP are EVKMDAEF and VHHQKLVF, respectively.   A series of fluorogenic substrates containing these peptide sequences were prepared using the automated solid phase synthesis procedure, as shown in Table III below (9, 49).   AMY.1 contains the octapeptide with Lys-Dabcyl at the C-terminus and Glu-Edans at the N-terminus, with free carboxylate and amino groups at the termini.   The amyloidogenic substrates AMY.2, AMY.3 and AMY.4 encompass the same Aβ N-terminal cleavage site Met-Asp within the octapeptide EVKMDAEF, and have C-terminal amide and N-terminal acetyl groups, but differ in the peripheral residues.   In substrate AMY.5, the two Glu residues within the octapeptide have been changed to Gln, in order to evaluate the role of negative charge at these positions.   Several secretory substrates, SEC.1 and SEC.2 were also prepared.   SEC.1 was analogous to AMY.1 with free carboxylate and amino termini, while SEC.2 was analogous to AMY.2.

Table III.   Fluorogenic substrates for Alzheimer's disease related proteases [a]

Secretory Cleavage Site

SEC.1	H_2N-Arg-J-Val-His-His-Gln-Lys-Leu-Val-Phe-B-Arg-CO_2H
SEC.2	Ac-Glu-J-Val-His-His-Gln-Lys-Leu-Val-Phe-B-Glu-NH_2
	(608-615, APP695)

Amyloidogenic N-Terminal Cleavage Site

AMY.1	H_2N-Arg-J-Glu-Val-Lys-Met-Asp-Ala-Glu-Phe-B-Arg-CO_2H
AMY.2	Ac-Glu-J-Glu-Val-Lys-Met-Asp-Ala-Glu-Phe-B-Glu-NH_2
	(593-600, APP695)
AMY.3	Ac-Arg-J-Glu-Val-Lys-Met-Asp-Ala-Glu-Phe-B-Arg-NH_2
AMY.4	Ac-Gln-J-Glu-Val-Lys-Met-Asp-Ala-Glu-Phe-B-Gln-NH_2
AMY.5	Ac-Gln-J-Gln-Val-Lys-Met-Asp-Ala-Gln-Phe-B-Gln-NH_2

[a]Standard single-letter or three-letter symbols for natural amino acids are used. Residue "J" denotes Glu(Edans) which has the fluorophore Edans attached to the side-chain carboxy group and residue "B" denotes Lys(Dabcyl) which has the chromophore Dabcyl attached to the side-chain amino group.

Incubation of substrates SEC.1 and AMY.1 with brain extracts resulted in extensive proteolysis by exopeptidases, in addition to any endopeptidase cleavages that may have occurred. These substrates were not employed in further studies of extracts containing exopeptidases.

An interesting report was published by Abraham and co-workers in 1991, describing a Ca^{2+}-activated protease activity potentially relevant to the N-terminal amyloidogenic cleavage (50). This activity had been characterized using a radiolabeled peptide with a sequence similar to AMY.1-4. We used these substrates to evaluate protease activities in brain tissue extracts, and evaluated the influence of Ca^{2+} on the proteolysis kinetics of the substrates. The proteolysis rate data shown in Table IV shows a wide variation for the different amyloidogenic substrates when incubated with soluble brain proteases at neutral pH, in the presence or absence of Ca^{2+}. The relative rates in the absence of Ca^{2+} showed a 6.8-fold variation from the least active (AMY.2, relative rate 1.0) to the most active (AMY.5, relative rate 6.8). Similarly, the effect of 1.0 mM Ca^{2+} ranged from nearly 4-fold stimulation for AMY.2 to only 1.5-fold stimulation for AMY.3 and AMY.5.

Table IV. Relative rates of amyloidogenic substrate proteolysis and Ca^{2+} effect [a]

Substrate	Net Charge	Relative Rate		Rate Increase
		No Ca^{2+}	1.0 mM Ca^{2+}	(x-fold) by Ca^{2+}
AMY. 2	-4	1.0	3.9	3.9
AMY. 3	0	4.0	6.1	1.5
AMY. 4	-2	1.3	3.9	3.0
AMY. 5	0	6.8	10.0	1.5

[a]The substrates were assayed using soluble fraction from normal brain tissue samples and the data represents the average values from analysis of two normal brain extracts. Proteinase activity assays were carried out in 40 mM Tris buffer (pH 7.5) using 10 µl of brain extract, with a total assay volume of 250 µl and a final substrate concentration of 10 µM, either in the absence of Ca^{2+} or in the presence of 1 mM Ca^{2+}. All rates were relative to the proteolysis rate of substrate 2.2 in the absence of Ca^{2+}.

Data for the closely analogous substrates (Table IV) permits dissection of factors that mediate the cleavage rate and the observed Ca^{2+} stimulation. Careful examination of the data indicates that both the cleavage rate and the extent of Ca^{2+} rate enhancement correlates with the number of negative charges. As a general rule, more negatively charged substrates were less reactive and exhibited larger rate enhancement by Ca^{2+}. Thus the Ca^{2+} stimulation appears to be more closely associated with interaction of Ca^{2+} with the substrate, rather than the proteases carrying out the cleavage.

Proteolytic Cleavage Sites. Proteolytic cleavage that occurs at any site between the fluorophore and chromophore can produce a fluorescence signal, so it is important to determine specifically the identity of the cleavage site or sites in instances where more than once protease activity may be present. This is accomplished readily by separation of substrate proteolysis fragments by HPLC and identification by electrospray ionization mass spectrometry (51). In the case of the APP-derived secretory substrate SEC.2, incubation with either normal or AD brain extracts at pH 7.5 produced one major endopeptidase cleavage after the Lys-7, generating Ac-Glu-Glu(Edans)-Val-His-His-Gln-Lys as the major N-terminal fragment. Over time, this fragment was processed further by carboxypeptidase activity to produce smaller peptides. The C-terminal fragment corresponding to this N-terminal fragment was Leu-Val-Phe-Lys(Dabcyl)-Glu-NH$_2$, but this fragment was not observed. The only detectable C-terminal fragment was Lys(Dabcyl)-Glu-NH$_2$, the end product of aminopeptidase digestion of Leu-Val-Phe-Lys(Dabcyl)-Glu-NH$_2$.

The amyloidogenic substrate AMY.2, upon proteolysis by normal or AD brain extract, also generated only one detectable endopeptidase cleavage, after Ala-8, producing one major N-terminal fragment Ac-Glu-Glu(Edans)-Glu-Val-Lys-Met-Asp-Ala. Two smaller fragments, Ac-Glu-J-Glu-Val-Lys-Met and Ac-Glu-J-Glu-Val-Lys, were also observed in amounts ranging from 30% to 100% of Ac-Glu-J-Glu-Val-Lys-Met-Asp-Ala. However, these two smaller peptide fragments were attributed to carboxypeptidase processing of Ac-Glu-J-Glu-Val-Lys-Met-Asp-Ala based on the time course of their production (51).

It was remarkable that both SEC.2 and AMY.2 apparently suffered only one initial endopeptidase cleavage, a possible indication that only one endopeptidase activity was responsible for the substrate cleavage. Regarding the cleavage of SEC.2 after lys-7, corresponding precisely to the secretory cleavage site observed in the APP-transfected HEK 293 cells, it should be noted that the cell culture secretase exhibited little sequence specificity and is probably a membrane bound enzyme cleaving at a certain distance beyond the membrane (47). Thus the soluble protease detected by SEC.2 in brain extracts is not likely to be the relevant cellular APP-secretase enzyme. The cleavage of the amyloidogenic substrate AMY.2 at pH 7.5 occurred at the carboxy side of Ala 8, a site inconsistent with the site corresponding to the reported N-terminus residue of Aβ peptide. One explanation for this result is that biologically relevant cleavage may occur at low pH, since evidence exists for the production of amyloidogenic APP fragments in endosomal or lysosomal compartments (52). It is possible that the Dabcyl or Edans groups influence enzyme-substrate interactions to bias the cleavage towards the Ala-8 site. These possibilities are being investigated.

Summary

Internally quenched fluorogenic substrate probes have been shown to be useful in a variety of contexts, from conventional enzymology and inhibitor evaluation, to searches for important new therapeutic protease targets. The development of automated methods for synthesis of these substrate probes has enabled access to a wide range of substrate sequences, and facilitated detailed studies of enzyme-substrate sub-site interactions. The selection of the Edans-Dabcyl fluorophore-quencher pair has led to the synthesis of efficiently quenched peptide substrates containing more than 30 amino acid residues. The overall versatility of these substrate probes ensures widespread future application to studies of many new and existing proteases of interest.

References

1. Finley, D.; Chau, V. *Ann. Rev. Cell Biol.* **1991**, *7*, 25.
2. Reid, I. A. *Arch. Inter. Med.* **1985**, *145*, 1475.
3. Yanagisawa, M.; Kurihara, H.; Kimura, S.; Tomobe, Y.; Kobayashi, M.; Mitsui.; Yazaki, Y.; Goto, K.; Masaki, T. *Nature,* **1988**, *332*, 411.
4. Rosenthal, P. J.; Mckerrow, J. H.; Aikawa, M.; Nagasawa, H.; Leech, J. H. *J. Clin. Invest.* **1988**, *82*, 1560.
5. Kohn, N. E.; Emini, E. A. ; Schleif, W. A.; *Proc. Natl. Acad. Sci. USA* **1988** *85,* 4686.
6. Lottenberg, R. et al. *Methods in Enzymology,* **1981**, *80,* 341.
7. Wang, G. T.; Huffaker, J. A.; Matayoshi, E.D.; Krafft, G. A. *Tetrahedron Lett.* ; **1990**, *31*, 6493.
8. Matayoshi, E. D.; Wang, G. T.; Krafft, G. A.; Erickson, J. W. *Science,* **1990**, *247*, 954.
9. Wang, G. T., Krafft, G. A. *Bioorg. Med. Chem. Lett.* **1992**, *2*, 1665.
10. Wang, G. T.; Chung, C. C.; Holzman, T. F.; Krafft, G. A. *Anal. Biochem.* **1993**, *210*, 351.
11. Claeson, G. ; Aurell, F.; Friberger, P.; Gustavsson, S.; Karlsson, G. *Haemostasis,* **1978**, *7*, 62.
12. Zimmerman, M.; Quigley, J. P.; Ashe, B.; Dorn, C.; Goldfarb, R.; Troll, W. *Proc. Natl. Acad. Sci. USA.* **1978**, *75*, 750.
13. Latt, S. A., Auld, D. S., Vallee, B. L. *Anal. Biochem..* **1972**, *50*, 56.
14. Kohn, N. E.; Emini, E. A.; Schleif, W. A.; et al. *Proc .Natl. Acad. Sci. USA,* **1988**, *85*, 4686.
15. Mous, J.; Heimer, E. P.; Le Grice, S. T.; *J. Virology,* **1988**, 1433.
16. Graves, M. C.; Lim, J. J.; Heimer, E. P.; Kramer, R. A. *Proc. Natl. Acad. Sci. USA.* **1988**, *85*, 2449.
17. Norbeck, D. W.; Kempf, D. J. *Ann. Rep. Med. Chem.* **1991**, *26*, 141.
18. Robins, T.; Plattner, J. J. *Acqu. Immunodef. Syndrome,* **1993**, *6*, 162.
19. Wlodawer, A.; Miller, M.; Jaskolski, M. *Science,* **1989**, *245*, 616.
20. Swain, A. L.; Miller, M. M.; Green, J.; Rich, D.; Schneider, J.; Kent, S. B.; Wlodawer, A. *Proc. Natl. Acad. Sci. USA,* **1990**, *87*, 8805.
21. Erickson, J. W.; Neidhart, D. J.; et. al. *Science,* **1990**, *249*, 527.
22. Kempf, D. J.; Norbeck, D. W.; Codacovi, L. M.; et al. *J. Med. Chem.* **1990**, *33*, 2687.
23. Kempf, D. J.; Sowin, T. J.; et al. *J. Org. Chem.* **1992**, *57*, 5692.
24. Kageyama, S.; Weinstein, J.; Shirasaka, T. et al. *Antimicrob. Agents Chemother.* **1992**, *36*, 926.
25. Darke, P. L.; Nutt, R. F.; Brady, S. F.; et al. *Biochem. Biophys. Res. Comm.* **1988**, *156*, 297.
26. Darke, P. L.; Leu, C-T.; Davis, L. J.; Heimbach, J. C.; Diehl, R. E.; Hill, W. S.; Dixon, R, A.; Sigal, I. S. *J. Biol. Chem.* **1989**, *264*, 2307.
27. Peng, C.; Ho, B. K.; Chang, T. W.; Chang, N. T. *J. Virology,* **1989**, *63*, 2550.
28. Pichuantes, S.; Babe, L. M.; Barr, P. J.; Craik, C. S.; *Proteins: Struc. Func. Genetics.* **1989**, *6*, 324.
29. Debouck, C.; Gorniak, J. G.; Strickler, J. E.; Meek. T. D.; Metcalf, B. W.; Rosenberg, M. *Proc. Natl. Acad. Sci. USA* **1987**, *84*, 8903.
30. Billich, S.; Knoop, M.-T.; Hansen, J.; Strop, P.; Sedlacek, J., Mertz, R.; Moelling, K. *J. Biol. Chem.* **1988**, *263*, 17905.
31. Richards, A.; Phylip, L. H.; Framerle, W. G.; Scarborough, F.; et al. *J. Biol. Chem.* **1990**, *265*, 7733.
32. Billich, S.; Winkler, G. *Peptide Res.* **1990**, *3*, 274.
33 Laragh, J. H.; *J. Hypertension,* **1986**, *4*, S143
34. Boger, J.; *Ann. Rep. Med. Chem.,* **1985**, *20*, 257.

35. Alderman, M. H.; et al. *New Engl. J. Med.*, **1991**, *324*, 1098.
36. Haber, E.; Koerner, T.; Page, L. B.; Kliman, B.; Purnode, A. *J. Clin. Endocrin.* **1969**, *29*, 1349.
37. Wathen, L. K.; Nuorala, K. W.; Wathen, M. W. *J. Clin. Lab. Anal.* **1991**, *5*, 282.
38. Reinhaez, A.; Roth, M. *Eur. J. Biochem.*, **1969**, *7*, 334.
39. Murakami, K.; Ohsawa, T.; Hirose, S.; Takada, K.; Sakakibara, S. *Ann. Biochem.*, **1981**, *110*, 232.
40. Poorman, R. A.; Palermo, D. P.; Post, L. E.; Murakami, K.; Kinner, J. H.; Smith, C. W.; Reardon, I.; Heinrikson, R. L. *Prot. Struc., Func., Genet.*, **1986**, *1*, 139.
41. Holzman, T. F.; Chung, C. C.; Edalji, R.; Egan, D. A.; Gubbins, E. J.; Reuter, A.; Howard, G.; Yang, L. K.; Pederson, T. M.; Krafft, G. A.; Wang, G. T.; *J. Prot. Chem.* **1990**, *9*, 663.
42. Holzman, T. F.; Chung, C. C.; Edalji, R.; Egan, D. A.; Martin, M.; Gubbins, E. G.; Krafft, G. A.; Wang, G. T.; Thomas, A. M.; Rosenberg, S. H.; Hutchins, C. *J. Prot. Chem.* **1991**, *10*, 553.
43 Robakis, N. K.; Ramakrishna, N.; Wolfe, G. and Wisniewski, H. M. *Proc. Natl. Acad. Sci. USA*, **1987**, *84*, 4190.
44. Kang, J.; Lemaire, H.-G.; Unterbeck, A.; Salbaum, J. M.; Masters, C. L. and Beyreuther, K. *Nature*, **1987**, *325*, 733.
45. Tanzi, R. E.; Guzella, J. R.; Watkins, P. C.; Bous, G. P.; George-Hyslop, P.; van Keuren, M. L.; Patterson, D.; Pagan, S.; Kurnit, D. M.; and Neve, R. L. *Science*, **1987**, *235*, 880.
46. Goldgaber, J.; Lerman, M. I.; McBridge, O. W.; Saffiotti, V. and Gajdusek, D. C. *Science*, **1987**, *235*, 877.
47. Sisodia, S. S.; Koo, E. K.; Beyreuther, K.; Unterbeck, A. and Price, D. L. *Science*, **1990**, *248*, 492.
48. Mori, H.; Takio, K.; Ogawara, M.; Selkoe, D. J. *J. Biol. Chem.* **1992**, *267*, 17082.
49. Wang, G. T.; Ladror, U. S.; Holzman, T F.; Klein, W. L.; Krafft, G. A. *Biochem. Biophys. Res. Comm.* Submitted.
50. Abraham, C. R.; Driscoll, J.; Potter, H.; van Nostrand, W. E. and Tempst, P. *Biochem. Biophys. Res. Comm.* **1991**, *174*, 790.
51. Ladror, U. S.; Wang, G. T.; Klein, W. L.; Krafft, G. A. and Holzman T. F. *J. Biol. Chem.* Submitted.
52. Golde, T. E.; Estus, S.; Younkin, L. H.; Selkoe, D. J. and Younkin, S. G. *Science*, **1992**, *255*, 728.

RECEIVED June 18, 1993

Chapter 13

Lifetime-Based Sensing Using Phase-Modulation Fluorometry

H. Szmacinski and J. R. Lakowicz

Department of Biological Chemistry, Center for Fluorescence Spectroscopy, University of Maryland School of Medicine, Baltimore, MD 21201

Fluorescence sensing is of widespread interest in the biophysical, biochemical and biomedical sciences (*1-10*) because of the potential for real-time sensing without the need for many steps associated with other analytical methods. Many fluorescent probes have been identified that change intensity in response to analytes of interest, and more have been synthesized for high specificity and/or sensitivity (*11*). Fluorescence is particularly well suited for optical sensing because emitted light returning from the sensor can easily be distinguished from the excited light. Owing to the sensitivity of fluorescence, even low fluorophore levels can be detected.

One can imagine fluorescence sensing based on intensity, intensity-ratio or lifetime measurements. These various methods for fluorescence sensing are illustrated in Figure 1. Intensity measurements have been used almost exclusively because of their simple instrumentation. Unfortunately, fluorescence intensity is sensitive to many other factors which can distort the observed values. The fluorescence intensity can vary due to excitation intensity, light losses in fibers, photobleaching or washout of the probes (concentration of dye is changing), as well as changes in the light scattering and/or absorption characteristics of the sample. These effects result in the need for frequent recalibration and other corrections. To avoid some of these problems wavelength-ratiometric probes have been developed, where ratio of signals at two excitation or emission wavelengths are measured. This technique is promising, but it has not found widespread use (except for calcium probes, fura-2 and indo-1), primarily due to the lack of suitable probes and known methods to create them. In some cases, fluorescence sensing relies on collisional quenching which does not cause a spectral shift of the quenched fluorophores. In fact, it appears that it will not be possible to create wavelength-ratiometric probes for collisional quenchers such as oxygen or chloride.

Because of advances in the technology for measuring lifetimes, particularly by the phase-modulation method (*12-17*), lifetime-based sensing (Figure 1) offers new opportunities for chemical sensing. This is because the fluorescence lifetimes of probes can be sensitive to a variety of factors or chemicals. Moreover, lifetime measurements are insensitive to probe concentration, photobleaching, washout and excitation instabilities. Another advantage of using lifetime-based sensing is the application to remote monitoring with fiber optics without the recalibrations and corrections necessary with intensity measurements. Rapid and continuous monitoring of many analytes (pH, pCO_2, O_2 etc.) is required in many areas of science, including chemistry, biochemistry, environmental sensing, clinical chemistry and industrial applications. The fluorescence lifetime or decay time can be determined from the slope of intensity following pulsed excitation,

$$I(t) = I_0 \exp(-t/\tau) \qquad (1)$$

where I_0 is the intensity at $t=0$ and τ is the decay time (Figure 1). For lifetime-based sensing one requires that the decay time (inverse slope in Figure 1, C) be dependent on the analyte of interest.

Instrumentation for Phase-Modulation Fluorometry

Fluorescence decay times are typically in the range of 1-20 ns, which requires short excitation pulses and high speed detectors and electronics (*18-19*). Simpler instrumentation can be imagined if the phase-modulation method is used. In this method, the pulsed source is replaced with an intensity-modulated source with modulation frequencies (f) ranging from 1-200 MHz. The information needed to determine the decay time is the phase angle shift (θ) of the emission relative to incident light, or the modulation of the emission (m) relative to that of incident light (Figure 1, D). The phase and modulation are separate measurements, each of which is related to the apparent phase (τ_θ) and modulation (τ_m) lifetimes by,

$$\tau_\theta = (1/2\pi f)\tan\theta, \qquad \tau_m = (1/2\pi f)(1/m^2 - 1)^{1/2} \qquad (2)$$

While the presently available phase-modulation fluorometers are rather large and expensive, this is primarily the result of their ability to operate over a wide range of frequencies and wavelengths (*12-15*). These capabilities are required to resolve complex and/or multi-exponential decay of fluorescence intensity, but such capabilities are not needed for sensing applications. If the probe characteristics are known, the instrument can be designed with a single wavelength light source and to operate at just one light modulation frequency appropriate for the chosen probe. The phase and modulation at a single frequency are easily measured in a second or less of data acquisition. This specialization can result in decreased complexity and cost as shown in Figure 2. This schematic shows a phase-modulation fluorometer with HeNe or laser diode source and an acousto-optic (AO) modulator. The AO modulators are not convenient if a wide range of frequencies are required, but are ideal if only a few frequencies are needed. Also, laser diode sources can allow even simpler instrumentation because they can be modulated by the driving current (*20-21*), and laser diode sources have already been used for frequency-domain

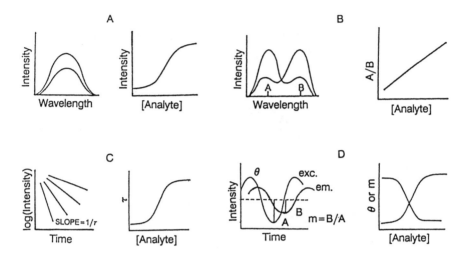

Figure 1.   Schemes for fluorescence sensing. A - single excitation or emission wavelength intensity, B - dual excitation or emission wavelengths intensity ratio; C - measurement of intensity decay; D - measurement of fluorescence phase angle and/or modulation.

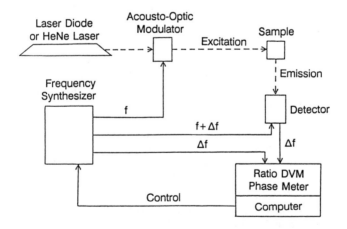

Figure 2.   A cross correlation phase-modulation fluorometer.

measurements to 2 GHz (*16-17*). However, the use of laser diode sources for lifetime-based sensing requires the development of red and/or NIR lifetime probes. The electronics for phase-modulation sensing can be based on relatively low-cost components (*22-23*).

An advantage of phase-modulation fluorometry for sensing is the use of cross-correlation detection, which results in an increased signal-to-noise ratio and rejection of undesired harmonics. The gain of the PMT detector is modulated by a small voltage applied to one of its dynodes. The gain modulation frequency is offset from the light modulation frequency by a small amount, typically 25 or 40 Hz. The phase and modulation information are preserved in the low frequency signal, and θ and m then easily measured with simple electronics. The AC component of the signal is typically passed through a 1 Hz wide filter, which rejects undesired harmonics. Cross-correlation electronics or numerical filtering results in phase and modulation measurements which are correct irrespective of the form of the light and gain modulation. The measurements do not require sinusoidal modulation, and in fact we frequently make use of pulse-sources for such measurements (*15*). The frequency synthesizers for a sensing application need not have a wide range, but the method does require that the two frequencies (f and f+Δf) be phase-locked. Finally, advances in probe chemistry are likely to result in lifetime-based probes which can be excited with laser-diode sources (*24-25*). This is advantageous because there is less autofluorescence with longer wavelength excitation, and it may become possible to perform non-invasive lifetime-based sensing through the skin. The skin becomes non-absorbing (scattering, but weakly absorbing) at wavelengths above 600 nm (*26*).

Lifetime-Sensing Using Collisional Quenching

One common approach to fluorescence sensing is to rely on fluorophores which are collisionally quenched by the analyte. There are many known collisional quenchers (analytes) which alter the fluorescence intensity and decay time. These include O_2 (*27-31*), chloride (*32-33*), chlorinated hydrocarbons (*34*), iodide (*35*), bromate (*36*), xenon (*37*), acrylamide (*38*), succinimide (*39*), sulfur dioxide (*40*), and halothane (*41*), to name a few. The quenching process obeys the Stern-Volmer equation:

$$I_0/I = 1 + K_{sv} [A], \qquad \tau_0/\tau = 1 + K_{sv} [A] \qquad (3)$$

where I_0, (τ_0), and I,(τ) are the fluorescence intensities (lifetimes), in the absence and presence of analyte (A), respectively. K_{sv} is the overall dynamic quenching constant which is equal to $k_d\tau_0$, with k_d being the diffusional bimolecular rate constant and τ_0 the unquenched lifetime of the indicator. In most cases the lifetimes and intensities decrease by the same amount, as indicated by equation 3. Because this type of quenching is a dynamic event involving the collision of dye and analyte, the process is sometimes called collisional quenching (in contrast to static quenching which is caused by the formation of stable ground state complex).

O$_2$ Sensing. Measurement of dissolved oxygen is important in a wide variety of situations, including blood gas measurement, monitoring of bioreactors in biotechnology applications, and in industrial process control. At present, most such measurements are performed using the Clark electrode (*42*). Oxygen optrodes offer many advantages over the amperometric Clark electrode. In contrast to a Clark electrode, an oxygen optrode does not consume oxygen, is not dependent on the oxygen flux, is electronically neutral, and can be of small size.

A variety of chemically quite different indicators and methods have been applied for probing oxygen, including hemoglobin absorbance (*43*), chemiluminescence (*44*), phosphorescence quenching (*45-47*), fluorescence quenching (*47-53*), and luminescence quenching of transition-metal complexes (*30,54-57*). Polycyclic aromatic hydrocarbons such as pyrene, pyrenebutyric acid, fluoranthene, decacyclene, and diphenylanthracene are viable indicators because they are efficiently quenched and fairly soluble in silicone (the most frequently used polymer). A variety of longwave absorbing dyes such as perylene dibutyrate, trypaflavin and porphyrins are known to be quenchable by oxygen when adsorbed on polar supports (silica gel). They have favorable absorptions and emissions, but their photostability is generally poor. A third group of indicators comprises metal-organic complexes of ruthenium, osmium, iridium, and platinum. These have strong metal-to-ligand energy transitions and long-lived excited states (up to 5 μs), which makes them useful for lifetime-based oxygen sensors. Most of these indicators are highly specific. Major interferents are sulphur dioxide, halothane, chlorine and some nitrogen oxides.

In order to obtain an oxygen sensor it is necessary to immobilize the oxygen-sensitive dye on a rigid support. A simple way of immobilizing a lipophilic indicator is to dissolve it in a hydrophobic polymer such as poly(vinyl chloride) (PVC) or silicone. Silicone, in particular, has excellent oxygen permeability and solubility, but is a poor solvent for most dyes. PVC, on the other hand is good solvent for most polycyclic aromatic hydrocarbons, but has slow oxygen diffusion. Silicones have excellent optical properties, can be handled easily, and allow the fabrication of extremely thin films. The observed quenching constants for a given indicator are highest in silicone, when compared with other polymer materials.

Practically all indicators are washed out by the sample unless covalently linked, resulting in a slow signal drift. The intensity can also be affected by photobleaching of the probe particularly under strong UV radiation. Probe wash-out, bleaching and lamp instability do not severely affect the performance of oxygen optrodes based on measurements of lifetime, because lifetime is independent of dye concentration (within reasonable ranges) as well as on changes in intensity signals due to lamp instability, detector sensitivity fluctuations, or light loses in the fibers.

For phase-modulation measurements we have chosen tris(4,7-diphenyl-1,10-phenanthroline)ruthenium(II) perchlorate ([Ru(Ph$_2$phn)$_3$]$^{2+}$), which is highly sensitive to oxygen. This complex in silicon is also rather specific. The physiologically and medically important gases, CO$_2$, nitrous oxide, cyclopropane, and halothane, are all without effect at concentrations well above those normally encountered. Also, the common petroleum industry pollutant H_2S as well as excellent solution quencher Fe^{3+} are without effect (*30*). Another advantage of this oxygen sensor is the relatively small dependence on temperature compared with other Ru(II) complexes (*30,58*).

Our experimental data are presented in Figure 3 using $[Ru(Ph_2phn)_3]^{2+}$ in propanol and in silicone. A HeCd laser with 442 nm excitation was used and measurements were carried out using frequency-domain instrumentation (*15*). In the typical oxygen range from 0 to 200 mm Hg, the phase changes 36 degrees in propanol and 25 degrees in silicone at 0.8 and 0.7 MHz modulation frequency (the lowest with our instrumentation). Using a lower modulation frequency of 0.2 MHz (dashed line) shows that oxygen can be measured with still higher accuracy (1-1.5 mm Hg).

The extent of collisional quenching is very sensitive to solvent viscosity, and also depends on temperature. Additionally, the unquenched $[Ru(Ph_2phn)_3]^{2+}$ lifetime in propanol decreases from 5.2 μs at 20 °C to 2.9 μs at 50 °C which results in decreased phase angle from 81.3 to 76.4 degrees at 0.2 MHz. This is a modest effect compared to the almost 70 degree change due to oxygen partial pressure from 0 to 200 mm Hg. Hence, temperature effects are simple to correct for, and in fact the unquenched lifetimes of other Ru(II) complexes can be used to measure the temperature (*30,58*, Lakowicz, J.R.; Szmacinski, H. unpublished observation).

Lifetime-Based Sensing Using Energy Transfer

Fluorescence lifetimes are also affected by fluorescence resonance energy transfer (FRET). The phenomenon of FRET is non-radiative energy transfer from the fluorescent donor to the acceptor, without emission and reabsorption of photons. The FRET is completely predictable based on the spectral properties of the donor and acceptor (*59-60*). If the donor and acceptor are at a distance R within the characteristic Förster distance (R_0) for energy transfer (optimal for sensing if $0.8\,R_0 \leq R \leq 1.3R_0$), and if the acceptor absorption spectrum changes in response to the analyte, then the lifetime of the donor will change in response to the analyte. Consequently, FRET can be used for sensing a wide variety of analytes (*61-66*).

CO_2 Sensing. Traditionally, gaseous CO_2 has been assayed via infrared absorption, or electrochemically by measuring changes in the pH of an inner buffer solution as a result of a varying CO_2 partial pressure above the solution. IR absorption also has been applied to optical sensors. CO_2 has a strong infrared absorption band extending from 4.2 to 4.4 μm (*67*). The IR approach is difficult to use with aqueous CO_2 solutions, so that the method based on pH coupling to dissolved carbon dioxide via a fluorescent sensing dye as the transducer has been adapted to optical techniques. It is well known that carbon dioxide will dissolve in water, according to the equation:

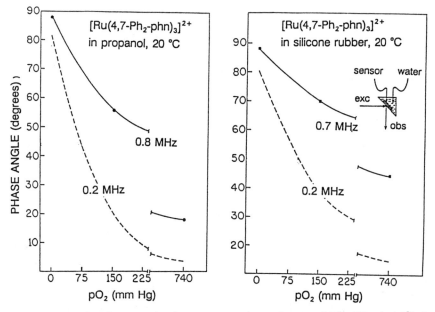

Figure 3.   O₂-phase angle fluorescence dependence of [Ru(Ph₂phn)₃]²⁺ in propanol (left) and in silicone (right) with excitation at 442 nm (HeCd laser). Dashed lines are expected data at optimal frequency for sensing (0.2 MHz) (not available in current instrumentation described in (15)).

$$CO_2(aq) + H_2O \rightleftharpoons H_2CO_3 \rightleftharpoons H^+ + HCO_3^-$$

At equilibrium, the pH in the internal solution depends on the concentration of carbonic acid produced upon hydration of the permeated CO_2, which in turn is proportional to the partial pressure of the analyte in the sample. Various pH-sensitive fluorescent probes have been used for measurements of carbon dioxide based on the pH change in an inner buffer solution caused by permeation of carbon dioxide through different gas-permeable membranes (68-74).

Decay time sensing of $pH(pCO_2)$ has been limited by the lack of pH probes which can be excited with simple laser sources like the HeNe or diode lasers. The fluorescence of the $pH(pCO_2)$ sensor must display good absorption of the laser light, a high quantum yield, and sensitivity in the desired pH range. It is difficult to obtain all of these requirements in a single chromophore. However, these characteristics are more easily obtained by using a two-part sensor, consisting of an energy transfer donor-acceptor pair. In this case, the fluorescent donor can be selected for its absorption, emission and decay characteristics, without concern for its sensitivity to $pH(pCO_2)$. The acceptor need not be fluorescent, and need only display a change in absorption in response to $pH(pCO_2)$ in the wavelength range of the donor emission. The donor and acceptor can be covalently linked or simply mixed together as described below.

The mechanism of inducing a $pH(pCO_2)$-dependent change in the donor decay times is a difference in efficiency of fluorescence resonance energy transfer from the donor to acid and base forms of the acceptor.

For unlinked donor-acceptor pairs the phenomenon of FRET requires an acceptor concentration in the range of 1-10 mM. Such acceptor concentrations result in high optical densities at the excitation and emission wavelengths, making intensity measurements difficult to use in a quantitative manner (75). At these high acceptor concentrations there are, in addition to FRET, inner filter effects which depend on the excitation and observation wavelengths and on the detailed macroscopic properties of the sample and detection optics. Since the phenomenon of FRET is predictable, this transduction mechanism can be extended to use with modulated laser diode sources by selection of alternative donor-acceptor pairs.

In our experiments the donors were selected to be excitable with a 543 nm HeNe laser. The donors and acceptor were dissolved at concentrations listed in Figure 4 in a 40 mM $NaHCO_3$ solution and mixed with a polyHEMA hydrogel (66). The equilibrated hydrogels were sandwiched between silicone rubber membranes, and sealed to retain the indicator solutions.

For testing the $pH(pCO_2)$ response of the sensors we used a 543 nm HeNe laser, for which the intensity was modulated with an acousto-optic modulator. This light source was chosen because it is practical for analytical or clinical sensing applications. Figure 4 shows CO_2-phase dependence for three donor-acceptor probes. The phase angles increased monotonically with increased partial pressures of CO_2, and the modulation decreased (not shown) indicating less FRET, resulting from the increased decay times of the donor. In these

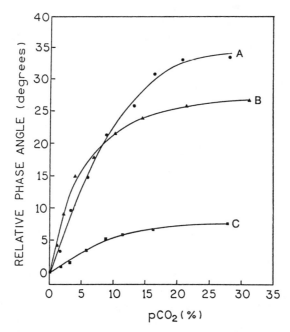

Figure 4. Dependence of donors fluorescence phase angles on pCO_2. The donor-acceptor pairs were in polyHEMA hydrogel, 25 °C, $\lambda_{exc} = 543$ nm. (A) Texas Red Hydrazide + Bromothymol Blue ($c_A = 0.002$ M), $\lambda_{obs} = 600$ nm, F = 133 MHz; (B) Eosin + Phenol Red ($c_A = 0.004$ M), $\lambda_{obs} = 580$ nm, F = 155 MHz; (C) Rhodamine 6G + Phenol Red ($c_A = 0.004$ M), $\lambda_{obs} = 600$ nm, F = 133 MHz.

sensors the amount of FRET decreases as the pH decreases (pCO_2 increases) due to an increased concentration of the acid form of the acceptor, which has a lower FRET efficiency than the base form of the acceptor.

There are two disadvantages of this type $pH(pCO_2)$ sensors using a mixture of donor and acceptor molecules. The first is attenuation of the observed intensity of the donor due to the partial absorption of the incident and emission light by the acceptor (concentration 1-10 mM). However, within some limits, an attenuation of intensity does not affect the decay time. A second disadvantage is the dependence of the extent of energy transfer on the acceptor concentration. It is quite possible that the acceptor concentration can change due to water evaporation, leakage, or due to osmotic effects. The changes in acceptor concentration will require recalibration of the $pH(pCO_2)$ sensor. The two discussed disadvantages can be avoided by the use of covalently linked donor-acceptor pairs. Because the acceptor is covalently linked to the donor, both are present in a one-to-one ratio, and the effective acceptor concentration does not depend on its bulk concentration. The long-wavelength absorption pH indicator Congo Red (*76*), may be suitable to create potential fluorescence pH probes on basis of energy transfer from a donor which can be excited at 633 nm (HeNe laser).

Glucose Sensing. A variety of methods have been suggested for optical measurements of glucose. The proposed methods include measurements of the oxygen consumption by glucose oxidase using an oxygen optrode (*77-78*), measurements of the changes in pH which accompany glucose oxidation (*79-80*), use of the intrinsic flavin fluorescence of glucose oxidase (*81*), and direct measurements of glucose by Fourier-transform infrared-spectrometry (FT-IR) (*82-83*). Others have proposed competitive displacement of fluorescently labeled Concanavalin A (ConA) from polymers by glucose (*62,84*). The association between ConA and dextrans was followed by removal of fluorescein-labeled ConA from the region of observation by binding to surface-bound ConA (*84*), or by energy transfer from the fluorescein-labeled dextran to rhodamine-labeled ConA (*62*). However, these glucose optrodes relied on measurements of the fluorescence intensity. While there is a growing understanding of the optical properties of tissues (*26,85*), it is still not practical to perform quantitative intensity measurements in highly scattering media. However, the increasing case of fluorescence lifetime-based sensing, creates new possibilities for non-invasive measurements of glucose.

A lifetime-based glucose assay, using the phenomenon of FRET, is illustrated in Figure 5. The assay consists of the donor covalently linked to ConA and sugar labeled with acceptor. Binding of sugar-acceptor to ConA-donor is expected to decrease the decay time of the donor fluorescence due to the FRET. Glucose in the sample will displace the sugar-acceptor, resulting in an increased lifetime which can be measured by the phase and/or modulation of the donor emission. It is important to recognize that the spectral properties of the donor and acceptor in an energy transfer-based assay can be adjusted as desired. FRET is a through space interaction, which is predictable based on the spectral properties of the donor and acceptor, and FRET is insensitive to the details of the local environment. Since the skin is non-absorbing at wavelengths above 600 nm, it may be possible to perform lifetime-based glucose measurements directly in tissues.

To illustrate the FRET-based glucose assay, we describe the results using a ConA labeled with 7-amino-4-methyl-coumarin (donor) and α-D-Mannoside labeled with tetramethylrhodamine cadaverine (TRITC-cadaverine) as the acceptor (Figure 6) (65). This donor-acceptor pair has a characteristic Förster distance of about 42 Å.

Single frequency phase and modulation assays are shown in Figure 7. The phase angle decreases (left) and modulation increases (right) upon addition of mannoside-TRITC because of increased energy transfer efficiency from donor to acceptor. The amount of energy transfer of the assay can be optimized by adjustment of the mannoside-TRITC concentration for optimal glucose concentration sensing. Addition of glucose to the sample results in reversal of the acceptor-induced changes in phase and modulation (Figure 7, inserts).

These results demonstrate that glucose assays based on FRET can be performed by phase-modulation measurements at single light-modulation frequency. Importantly, the energy transfer mechanism can be confidently expected to work at all wavelengths, so that an assay can be designed which takes advantage of currently available long-wavelength probes, lasers and detectors.

Lifetime-Based Sensing Using Intrinsic and Conjugated Probes

Another class of fluorescence sensors are those molecules which display changes in lifetime in response to the analyte of interest. Such sensors are distinct from collisionally quenched sensors, such as the ruthenium complexes for oxygen, in that the mechanism is not necessarily collisional quenching. The lifetimes of such sensors can increase or decrease in response to analyte.

One can imagine two classes of these lifetime sensors. We refer to these as intrinsic and conjugate sensors. A conjugate sensor is one consisting of a fluorophore covalently linked, via an inert bridge, to the analyte binding position of the sensor. An intrinsic sensor is one in which the analyte-sensitive region is built into the fluorophore. We are already aware of intrinsic probes for measurement of pH (86) and intrinsic and conjugated probes for Ca^{2+} (87-91) and Mg^{2+} (92). To illustrate the different types of sensors, we regard indo-1 and quin-2 as intrinsic Ca^{2+} sensors, and Calcium Green as a conjugate Ca^{2+} sensor (Figure 8).

pH Sensing. Rapid and continuous monitoring of pH is required in many areas of science. The glass electrode is most widely used because it is reliable and measurements can be done rapidly. In clinical applications, where monitoring of blood pH is desirable during complex surgical procedures and intensive care, the use of glass electrodes mounted on flexible catheters were found to be impractical because of their size, rigid design, and potential electrical hazards.

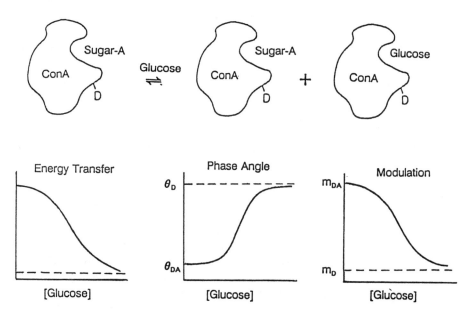

Figure 5. Intuitive description of an energy transfer-based glucose assay. Displacement of sugar-acceptor from ConA-donor by glucose decreases FRET, resulting in phase angle increase and modulation decrease of the donor fluorescence.

Donor: ConA-AMCA

Acceptor: TRITC-Mannoside

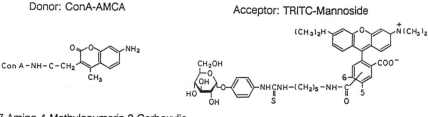

7-Amino-4-Methylcoumarin 3-Carboxylic
Acid Succinimidyl Ester (AMCA)
Concanavalin A (ConA)

Tetramethylrhodamine Cadaverine (TRITC)
a,D-Mannose Pyranosyl Phenyl (Mannoside)

Figure 6. Fluorescence donor and acceptor for energy transfer-based glucose assay.

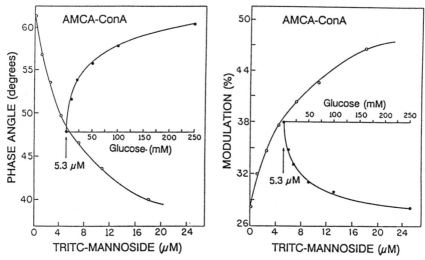

Figure 7.   Phase fluorescence (left) and modulation (right) assay of glucose based on energy transfer using AMCA-ConA and TRITC-Mannoside, $\lambda_{exc} = 360$ nm, 25 °C, F = 201.14 MHz. The inserts show the reversal of energy transfer by added glucose.

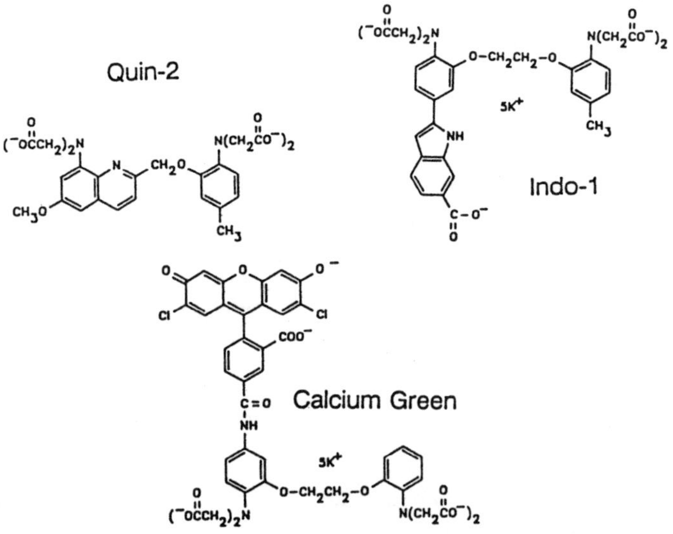

Figure 8.   Chemical structures of intrinsic probes (quin-2 and indo-1) and conjugated probe (Calcium Green).

Moreover, local electrical potentials are known to adversely affect the performance of pH electrodes.

Many of these problems can be overcome by optical measurements of pH. The most popular optical methods are those which rely on pH-dependence of absorption or fluorescence. A variety of such pH indicators are already known (*93*), but only a few meet the requirements of an excitation maximum beyond 400 nm to allow use of inexpensive plastic fiber optics as light guides, as well as inexpensive light sources such as halogen lamps, HeNe and the diode lasers. A popular family of pH indicators has been the fluoresceins, for which the fluorescence intensity is strongly pH dependent (*94-98*). Upon acidification, the emission spectra of all the fluorescein dyes decreases in amplitude without a significant wavelength shift, so that ratio-wavelength measurements are not possible. The excitation wavelength-ratiometric HPTS (8-hydroxypyrene-1,3,6-trisulfonate) on different solid supports has been investigated extensively (*49,99-100*). HPTS can be excited with visible wavelengths at 405 and 465 nm with an excellent ratio response to pH. Also, the use of two excitation wavelengths is not convenient in many applications, such as laser scanning microscopy and flow cytometry. Also, the loading of HPTS into living cells requires a permeabilization procedure. Review of fluorescent pH probes which fluorescence spectra are sensitive to pH and useful in biological applications has been recently published (*101*).

A more recent class of pH probes are the seminaphtofluoresceins (SNAFL) and the seminaphtorhodafluors (SNARF). These probes were reported for use as wavelength-ratiometric probes (*86*). Their absorption and emission spectra change significantly in response to pH.

We recognized that these pH sensors could be excited with conveniently simple lasers, such as the 543 nm HeNe laser. Additionally, we suspected that the spectral shifts could be accompanied by a change in lifetime. Hence, we measured the pH-dependent fluorescence decay times of the few SNAFL and SNARF probes. The detailed chemical and spectral properties of these probes are in (*86*).

The lifetimes were found to be strongly pH-dependent, making these probes useful as lifetime-based pH sensors. The acid and base forms have distinct lifetimes. The two representative frequency responses of the intensity decays are shown in Figure 9. Some of these probes behaves as carboxy SNARF-2 (acid form has a shorter lifetime) and some of them as carboxy SNARF-6 (acid form has a longer lifetime than base form, for detailed results see Table I). The frequency responses over the wide range of frequencies are necessary to analyze intensity decays and to choose an optimal frequency for sensing in regard of lifetimes values and heterogeneity. The optimal frequency can be regarded as this frequency at which there are maximum changes in phase and modulation in response to analyte. For SNAFL and SNARF probes there is a relatively wide range of frequencies with similar changes in phase and modulation (Figure 9). All of these probes can be excited with an inexpensive excitation source such as HeNe laser with 543 nm in which light can be modulated with an acousto-optic modulator. The pH-dependent phase angle of

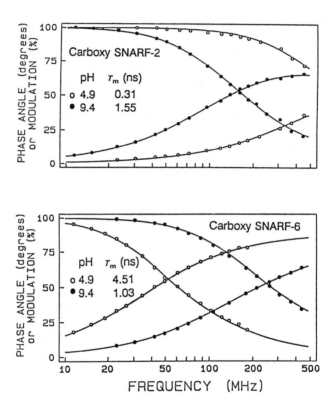

Figure 9. Frequency-responses of the intensity decays for acid and base forms of carboxy SNARF-2 (top) and carboxy SNARF-6 (bottom) with excitation at 563 nm and observation at 600 nm in 80 mM TRIS at 25 °C.

Table I.   **Lifetimes, phase and modulation characteristics of the pH indicators**

pH indicator	τ(ns)		Phase and modul.[a]		Apparent pK_a[b]
	Acid	Base	Δp (deg)	Δm (%)	
SNAFL-1	3.74	1.19	10.5	22.5	6.8
Carboxy SNAFL-1	3.67	1.11	23.0	35.0	6.3
Carboxy SNAFL-2	4.59	0.94	34.5	48.0	6.5
Carboxy SNAFL-2	---	---	---	---	(6.1-8.9)[c]
Carboxy SNARF-1	0.52	1.51	27.5	23.0	6.85
Carboxy SNARF-2	0.29	1.55	35.0	27.0	6.8
Carboxy SNARF-6	4.51	1.03	34.0	48.5	7.35
Carboxy SNARF-6	---	---	---	---	(6.8-9.5)[c]
Carboxy SNARF-X	1.79	2.59	12.0	16.5	7.35
BCECF[d]	3.17	4.49	10.0	14.0	7.35

[a] F_{mod} = 135 MHz, λ_{exc} = 543 nm (HeNe laser), λ_{obs} > 575 nm, 80 mM TRIS, T = 20 °C

[b] average value from phase and modulation calibration curves

[c] λ_{obs} from 560 to 660 nm with 10 nm width

[d] F_{mod} = 65 MHz, λ_{exc} = 442 nm (HeCd laser), λ_{obs} = 500 nm

several of these probes are shown in Figure 10. The increases in phase angle is related to an increase in mean lifetime (Equation 2). Most of these probes displayed changes in phase angle of more than 25 degrees, making them excellent lifetime-based pH probes. Similar changes or larger occur in the modulation (see Table I). The sensitive pH range of these probes can be dramatically shifted by a technically easy change in the observed emission wavelength. This is shown for the phase of carboxy SNARF-6 in Figure 11. The pH sensitive range of carboxy SNARF-6 excited at 543 nm is from 5 to 10.5. The apparent pKa (from phase or modulation plot) is dependent on the observation wavelength, and can be lower or higher than from the steady state intensity measurements. This wide pH range is possible because of the large wavelength shift between the emission spectra of acid and base forms (Figure 11, top). A similar effect can be achieved using different excitation wavelengths (not preferred with laser excitation sources). This lifetime-based property is highly desirable, particularly when in an environment of the high and low pH such as in cellular measurements (acid and base organelles), as well as in applications where a wide (over 2 pH units) range is desirable which cannot be covered by a single probe using intensity methods. One should note that sensitivity stays the same over the entire pH range (the same curve slopes at different observation wavelengths). The mean lifetimes, phase and modulation characteristics, and apparent pK_a are summarized in Table I. Additional details can be found in (Szmacinski, H.; Lakowicz, J.R. *Anal. Chem.*, 1993, in press). Some of these probes are available conjugated with dextran, and can be used behind dialysis membrane with fiber optic light guides (102). The results for BCECF (2',7'-bis(2-carboxyethyl)-5(and6)-carboxyfluorescein) are also included in Figure 10 and in Table I because this probe has been frequently used for estimation of intracellular pH (101), but it is likely that BCECF will be replaced with more convenient SNAFL and SNARF probes (dual wavelength-ratiometric and lifetime-based possibilities with wide range of apparent pK_a).

Calcium Sensing. Measurements of intracellular concentrations of Ca^{2+} are of wide interest in cell physiology and biology. There is a vast literature concerning the measurements of the intracellular concentrations of Ca^{2+} using fluorescent indicators and the journal Cell Calcium is devoted to this topic (11,101,103).

A number of fluorescent indicators are currently used, including quin-2 (104-105), fura-2 and indo-1 (106), fluo-3 and rhod-2 (107). The newest calcium probes, Calcium Green, Calcium Orange and Calcium Crimson have only been used in a few reports (89,108). These probes do not display shifts in the excitation or emission wavelengths and are thus difficult to use in microscopy and flow cytometry. At present, most measurements of intracellular calcium are performed using wavelength-ratiometric indicators such as fura-2 (dual excitation wavelength) and indo-1 (dual emission wavelength). This is because the concentration of the probe within the cell is not known and cannot be easily controlled. Furthermore, even if the probe concentration is known, the intensity measurements are not reliable due to photobleaching of the probe (particularly under microscopic illumination). At present, because of its higher intensity and

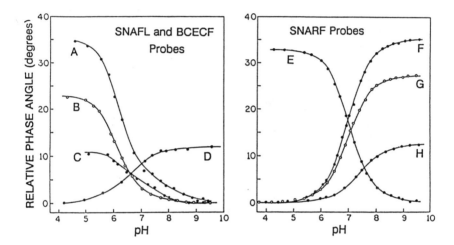

Figure 10.   pH-dependent phase angles for SNAFL and BCECF (left) and for SNARF probes (right) with excitation at 543 nm (HeNe laser), in 80 mM TRIS, 25 °C, $\lambda_{obs} > 575$ nm, F = 135 MHz. (A) carboxy SNAFL-2; (B) carboxy SNAFL-1; (C) SNAFL-1; (D) BCECF, $\lambda_{exc} = 442$ nm (HeCd laser), $\lambda_{obs} = 540$ nm, F = 65 MHz; (E) carboxy SNARF-6; (F) carboxy SNARF-2; (G) carboxy SNARF-1; (H) Carboxy SNARF-X.

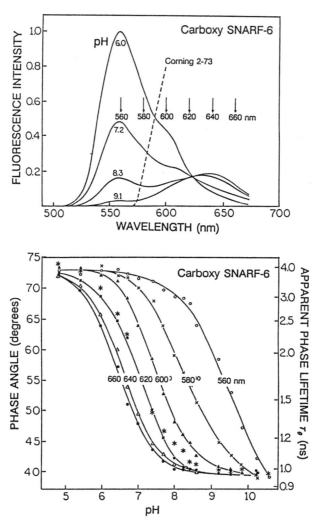

Figure 11. Emission spectra at various pH for carboxy SNARF-6 (top), arrows indicate the observation wavelengths with 10 nm width. pH- and emission wavelength-dependent phase angles for carboxy SNARF-6 (bottom). The stars are data using wideband filter Corning 2-73 ($\lambda_{obs} > 575$ nm)

photostability, fura-2 is now more widely used then any other of the calcium probes. It is regarded as an excellent excitation-ratiometric probe, with a large dynamic range of ratio signals (*106*). The probes fluo-3, rhod-2, and the newest Calcium Color series which can be excited with visible wavelengths, are less used because no spectra shift occurs upon calcium binding for neither excitation nor emission.   Quin-2, fura-2 and indo-1 require UV excitation , which excites significant autofluorescence from cells. UV excitation could cause biological side effects and penetrates microscope optics poorly.

We have characterized some of the available calcium probes as potential lifetime probes using phase-modulation fluorometry. The lifetime characteristics and phase-modulation calcium dynamic ranges are summarized in Table II and in Figure 12.

The lifetime properties of quin-2 have been reported recently by several groups, and the data are in agreement (*87-88,90-91*). Quin-2 undergoes the largest change in lifetime, displaying a change in phase angle of about 50 degrees, making it an extremely lifetime sensitive Ca^{2+} probe. For excitation at 342 nm (339 nm with intensity measurements), the apparent dissociation constant from phase-modulation measurements were 29 and 10 nM (see Table II), compare to 60 nM (*103*) or 48 nM (*88*) from intensity measurements. This shift of K_d is due to the almost 5-fold increase of quantum yield of the Ca-bound form (from 0.03 to 0.14 (*104*)) which has a long lifetime (11.6 ns). One can expect that longer excitation wavelengths could yield higher apparent K_d, which are more useful for intracellular calcium concentration measurements. The disadvantage of quin-2 is rather fast photobleaching compared to fura-2 and indo-1. Our investigations of quin-2 photoeffects revealed that during illumination quin-2 undergoes phototransformation as well as photobleaching depending on calcium concentration, with a significant change in lifetime (Lakowicz, J.R., et al. *Cell Calcium* 1993, in press). The phototransformation is a very undesirable effect and can cause problems with proper calibration curves regardless of the measurement method. Thus, quin-2 should be regarded as an excellent calcium lifetime-based probe with cell suspensions where photochemical effects can be avoided.

We have found that the mean lifetime of fura-2, the most widely used Ca^{2+} probe, changes only about 50% (from 1.09 ns to 1.68 ns) upon the Ca^{2+} binding. Similar values have been found by Keating and Wensel (*109*), the only other publication with lifetime measurements of fura-2. This small increase in lifetime is associated with a phase change of about 14 degrees (Figure 13). Because of the shift in the excitation spectra for free and Ca^{2+}-bound forms of fura-2, the apparent K_d is dependent on excitation wavelength. Using excitation wavelengths from 345 to 380 nm the dissociation constant can be varied from 60 to 1900 nM, increasing the range of measurable $[Ca^{2+}]$ up to 10 μM.

One may question the use of fura-2 as a lifetime-based calcium indicator because of the relatively small change in lifetime. An accuracy of 0.3° in phase angle at the steepest part of curves (Figure 13) will yield $[Ca^{2+}]$ values, which are accurate ± 9 nM at 345 nm, ± 30 nM at 365 nm and ± 130 nM at 380 nm excitation.   These results show that even with moderate changes in lifetime,

Table II.   **Lifetimes and phase-modulation characteristics of the calcium probes**

Indicator	τ (ns)		Phase and modul.		Apparent K_d		Experimental condition		
	Free anion	Ca-bound	Δp (deg)	Δm (%)	K_d^P (nM)	K_d^m (nM)	λ_{exc} (nm)	λ_{obs} (nm)	F (MHz)
quin-2	1.35	11.6	49.5	51.5	29	10	342	>450	49.34
fura-2	1.09	1.68	14.0	15.5	108	67	345	>450	208.73
fura-2	---	---	---	---	1900	1600	380	>450	208.73
indo-1	1.40	1.66	20.0	9.5	1130	953	345	>500	208.73
fura-red	0.12	0.11	3.5	4.0	---	---	567	640	1400.4
C. Green	1.05	3.59	48.5	49.0	41.5	13.5	514	>515	151.8
C. Orange	1.24	2.33	21.0	22.0	270	150	565	>575	121.44
C. Crimson	2.56	4.10	17.5	17.5	160	100	590	>610	102.46

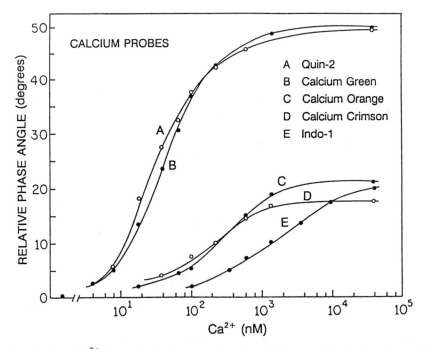

Figure 12. Ca^{2+}-dependent phase angles for calcium probes. Experimental condition are in Table II.

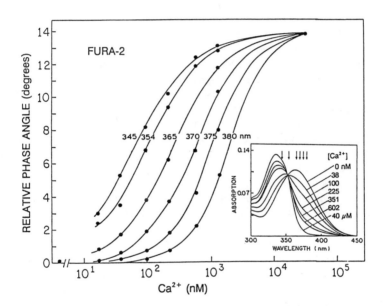

Figure 13.   Ca²⁺- and excitation wavelength-dependent phase angle for fura-2. The insert shows absorption spectra at various concentrations of Ca²⁺, arrows indicate the excitation wavelengths.

lifetime-based methods (more precisely, the phase-modulation method) can provide accurate measurements of intracellular calcium.

Indo-1 displays a similar range of phase angle changes in response to calcium (Figure 12), but it is necessary to choose the long-wavelength part of emission spectra. This effect is due to an excited-state reaction which is calcium-dependent (Szmacinski, H., et al. *Photochem. Photobiol.* 1993, in press). We note that the presence of an excited-state reaction could result in difficulties in the use of indo-1 as a lifetime or ratiometric calcium probe. This is because the rate and extent of the reaction may depend on the viscosity, ionic strength, pH and other additional conditions in the immediate environment of the fluorophore. An advantage of indo-1 is that it can be used with two-photon excitation (Szmacinski, H., et al. *Photochem. Photobiol.* 1993, in press) whereas we were unable to obtain useful two-photon induced fluorescence with quin-2 and fura-2.

In order to circumvent the need for UV excitation, a new series of calcium probes has been developed by Molecular Probes (*11*). These probes have absorption maxima at 508, 552, and 590 nm for Calcium Green, Calcium Orange and Calcium Crimson, respectively. There are no shifts in their absorption and emission spectra in response to Ca^{2+} (Figure 14). Consequently, these probes cannot be used as ratiometric indicators. However, the fluorescence intensities increase significantly in response to Ca^{2+}. We note that presence of detectable emission in the absence of Ca^{2+} is particularly suitable for lifetime-based sensing because the free anion forms display easily detectable emissions. Lifetime-based sensing requires that both forms (free anion and Ca^{2+}-complex) need to have contributions to the fluorescence signal. It is important to recognize that observations of Ca^{2+}-dependent changes in intensity do not necessarily imply a similar lifetime change. We observed mostly a smaller change in mean lifetime than in quantum yield (except of quin-2). The fluorescence intensity of Fura Red decreases upon Ca^{2+} binding, whereas the lifetime does not change (0.12 ns, Table II). For this reason it is important to characterize the intensity decay of the probes, over a range of Ca^{2+} concentrations. A detailed examination of the intensity decays has been already published for quin-2 (*88*), for Calcium Green, Orange and Crimson (*89*), for indo-1 (Szmacinski, H., et al. *Photochem. Photobiol.* 1993, in press) and for fura-2 (Lakowicz, J.R.; Szmacinski, H. prepared for publication). The intensity decays analyses can reveal lifetime components associated with free anion and Ca^{2+}-complex and the effect of photobleaching on these components.

The Ca^{2+}-dependent phase angles for Calcium Green, Calcium Orange and Calcium Crimson are shown in Figure 12. Calcium Green displays similar changes in phase and modulation as does quin-2. The apparent dissociation constants are also lower than those from intensity (*11,89,108*) (Table II). This is a result of the longer lifetime component (Ca-complex) with the ~9-fold higher quantum yield than the free anion. Other probes (Calcium Orange and Calcium Crimson) have smaller but still significant changes in phase and modulation with higher dissociation constants (Table II). All of the probes presented in Table II can be considered as lifetime-based calcium probes. The visible-wavelength probes without ratio-wavelength capabilities are especially promising.

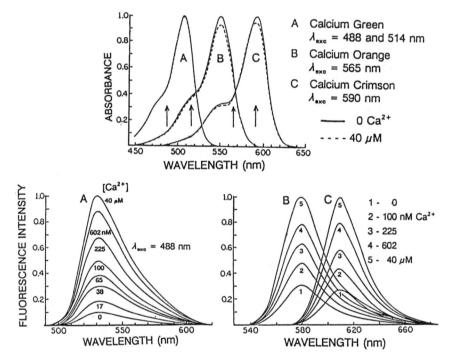

Figure 14.   Absorption (top) and emission (bottom) spectra for Calcium Green, Calcium Orange and Calcium Crimson at various calcium concentrations.

Conclusions

It seems valuable to speculate on the potential future applications of lifetime-based sensing. In our opinion, the dominant advantages of lifetime-based sensing are that the measurements can be independent of the local probe concentration, and the decay time measurements are mostly self-calibrating. The insensitivity of the lifetime to the local probe concentration enables the use of lifetime-based sensing in flow cytometry (110), fluorescence microscopy (25,88,111-112, Lakowicz, J.R. et al. *Cell Calcium*, in press), and clinical chemistry (25,65-66,113, Lakowicz, J.R. et al. *SPIE Proceedings* 1993, Vol.1895, in press)

The advantages of lifetime-based sensing are illuminated by the use of fluorescence in flow cytometry. The cells flow one-by-one through an observation area which is illustrated with a laser light source (Figure 15). It is difficult to make quantitive use of the fluorescence intensity because of the variations in cell volume and extent of probe uptake. In order to compensate for these intensity variations, there have been attempts to use wavelength-ratiometric probes. However, flow cytometry requires an emission wavelength-ratiometric probe since it is not practical to illuminate the cells with two different laser wavelengths, which must occur at different times or with temporal or spatial resolution. Because of these constraints there is currently only one Ca^{2+} probe useful in flow cytometry, this being indo-1 which requires UV excitation. Additionally, the intensity-based measurements are relative, not absolute, and require frequent recalibration.

Many of these difficulties can be circumvented by the use of lifetime-based sensing. More specifically, it is now possible to measure the phase angles on a cell-by-cell basis as they pass through the laser beam (110). This is accomplished using a RF circuit which "calculates" the phase angle of the microsecond light pulses from the cells as they pass through the laser beam; phase flow cytometry requires an intensity-modulated source. This modulation is readily accomplished using a commercially available acousto-optic modulator. The circuits for "calculation" phase angle are technologically simple (110), and can be readily constructed and inserted into standard flow cytometric pulse-processing electronics.

Addition of phase measuring capabilities to a flow cytometer immediately enables measurements of intracellular Ca^{2+} using a visible wavelength source. The probe Calcium Green displays a substantial change in phase angle in response to Ca^{2+} (89), and can be excited at 488 or 514 nm. At present, the phase angles measured in a flow cytometer (110) are already known to be independent of intensity over a 50-range of intensities, and higher dynamic ranges can be expected in the next generation circuits. This illustrates the self-calibrating feature of lifetime-based sensing, as the phase angle will represent the Ca^{2+} concentration, independent of the cell volume or extent of probe uptake by the cells.

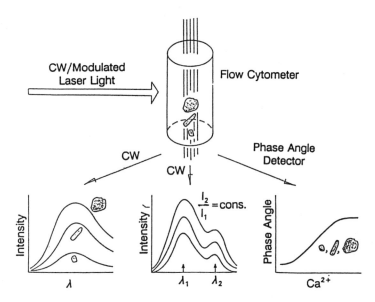

Figure 15.   Flow cytometer sorts cells based upon size. Differentiation can be by fluorescence intensity (left), wavelength ratios (middle), or phase angle detection (right).

Lifetime-based sensing also offers opportunities in fluorescence microscopy and other imaging applications, such as remote sensing of temperature and air flow. In fluorescence microscopy one generally cannot control the local probe concentration, and the intensity often changes rapidly due to photobleaching. Hence, the intensity image (Figure 16, lower left) is typically a probe concentration image, which is only of interest if the probe co-localizes with the analyte of interest. Fluorescence lifetime imaging, or FLIM, allows image contrast to be created based on the lifetimes of the probes at each point in the image (Figure 16, lower right). If the probe lifetime is sensitive to the analyte, one now has an ability to create chemical or biochemical images. As discussed above, probes are known which display lifetime changes to a wide range of analytes, enabling imaging of pH, O_2, Cl^-, Ca^{2+}, Mg^{2+}, and many others.

Until recently, there were no practical methods for measuring the numerous lifetimes needed for a moderate resolution image, such as 512 x 512 pixels. We have now developed a method which allows lifetime-sensitive information to be collected simultaneous at all pixels in the image (*110*). This method uses a gain-modulated image intensifier as an optical phase-sensitive imaging detector. The intensifier screen output is imaged onto a slow-scan CCD camera, which is highly desirable for most imaging applications. This device has already been used to image Ca^{2+} in cells (*25*, Lakowicz, J.R. et al. *Cell Calcium* 1993, in press), and to resolve free and protein-bound NADH (*112*).

In summary, we envision numerous applications for lifetime-based sensing not only in analytical chemistry, but also in the fields of cell biology, cell physiology, and remote-sensing applications.

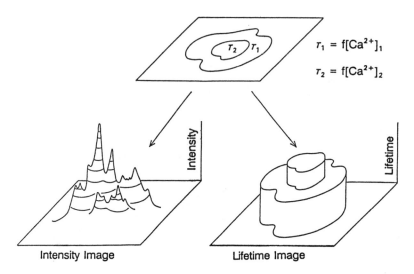

$\tau_1 = f[Ca^{2+}]_1$

$\tau_2 = f[Ca^{2+}]_2$

Intensity Image   Lifetime Image

Figure 16.   Intuitive description of fluorescence lifetime imaging (FLIM).

Aknowledgments

This work was supported by grants to the Center for Fluorescence Spectroscopy from the NSF (DIR-871040) and the NIH (RR-08119), with support for instrumentals from NIH-RR-07510. The authors also express appreciates from support the Medical Biotechnology Center at the University of Maryland at Baltimore.

References

1.   *Applications of Fluorescence in the Biomedical Sciences*; Taylor, D.L.; Murphy, R.F.; Lanni, F.; Birge, R.R., Eds.; Liss: New York, NY, 1986.
2.   *Chemical Sensors and Microinstrumentation*; Murray, R.W.; Dessy, R.E.; Heineman, W.R.; Janata, J.; Seitz, W.R., Eds.; ACS Symposium Series; American Chemical Society: Washington, DC, 1989; Vol. 403.
3.   *Luminescence applications in biological, chemical environment and hydrological sciences*; Golgberg, M.C., Ed.; ACS Symposium Series; American Chemical Society: Washington, DC, 1989; Vol. 383.
4.   *Fiber-Optic Chemical Sensors and Biosensors*; Wolfbeis, O.S., Ed.; CRC Press: Boca Raton, Fl, 1991; Vols. I and II.
5.   *Biophysical and biochemical aspects of fluorescence spectroscopy*; Dewey, T.G., Ed.; Plenum Press: New York, NY, 1991.
6.   *Luminescence techniques in chemical and biomedical analysis*; Baeyens, W.R.G.; De Keukelaire D.; Korkidis, K., Eds.; Marcel Dekker: New York, NY, 1991.
7.   *Fluorescence Spectroscopy, New Methods and Applications*; Wolfbeis, O.S., Ed.; Springer-Verlag: Berlin Heidelberg, 1993.

8.   Wolfbeis, O.S. *Trends Anal. Chem.* 1985, *4(7)*,184-188.
9.   Wolfbeis, O.S. *Pure Appl. Chem.* 1987, *59(5)*, 663-672.
10. Wolfbeis, O.S. *Fresenius Z. Anal. Chem.* 1986, *325*, 387-392.
11. Haugland, R.P. *Handbook of Fluorescent Probes and Research Chemicals*; Molecular Probes, Inc.: Eugene, OR, 1992.
12. Gratton, E.; Limkeman, M. *Biophys. J.* 1983, *44*, 315-324.
13. Lakowicz, J.R.; Maliwal, B. *Biophys. Chem.* 1985, *21*, 61-78.
14. Feddersen, B.A.; Piston, D.W.; Gratton, E. *Rev. Sci. Instrum.* 1989, *60*, 2929-2936.
15. Laczko, G.; Gryczynski, I.; Gryczynski, Z.; Wiczk, W.; Malak, H.; Lakowicz, J.R. *Rev. Sci. Instrum.* 1990, *61*, 2331-2337.
16. Berndt, K.W.; Gryczynski, I.; Lakowicz, J.R. *Rev. Sci. Instrum.* 1990, *61*, 1816-1820.
17. Thompson, R.B.; Frisoli, J.; Lakowicz, J.R. *Anal. Chem.* 1992, *64*, 2075-2078.
18. J.N. Demas, *Excited State Lifetime Measurements*, Academic Press 1983.
19. D.V. O'Connor and D. Phillips, *Time-Correlated Single Photon Counting*, Academic Press 1984.
20. K. Otsuka, Gigabit Optical Pulse Generation in Integrated Lasers and Applications, In *Picosecond Optoelectronic Devices*; Lee, C.H., Ed.; Academic Press, Inc. New York 1984.
21. A. Yariv, *Quantum Electronics*, John Wiley & Sons, pp 255-260, 1989.
22. Barbieri, B.; DePiccoli, F.; Gratton, E. *Rev. Sci. Instrum.* 1989, *60(10)*, 3201-3206.
23. Berndt, K.W.; Lakowicz, J.R. *Analyt. Biochem.* 1992, *201*, 319-325.
24. Mujumdar, R.B.; Ernst, L.A.; Mujumdar, S.R.; Wagooner, A.S. *Cytometry* 1989, *10*, 11-19.
25. Lakowicz, J.R.; Johnson, M.L.; Lederer, W.J.; Szmacinski, H.; Nowaczyk, K.; Malak, H.; Berndt, K. *Laser Focus World* May 1992,
26. Chance, B.; Leigh, J.S.; Miyake, H.; Smith, D.S.; Nioka, S.; Greenfield, R.; Finander, R.; Kaufman, K.; Levy, W.; Young, M.; Cohen, P.; Yoshioka, H.; Boretsky, R. *Proc. Natl. Acad. Sci. USA* 1988, *85*, 4971-4975.
27. Kautsky, H. *Trans. Faraday Soc.* 1939, *35*, 216-219.
28. Ware, W.R. *J. Phys. Chem.* 1962, *66*, 455-458.
29. Lakowicz, J.R.; Weber, G. *Biochemistry* 1973, *12*, 4161-4170.
30. Wolfbeis, O.S.; Urbano, E. *Fresenius Z. Anal. Chem.* 1983, *314*, 577-581.
31. Bacon, J.R.; Demas, J.N. *Anal. Chem.* 1987, *59*, 2780-2785.
32. Verkman, A.S.; Sellers, M.C.; Chao, A.C.; Leung, T.; Ketcham, R. *Analyt. Biochem.* 1989, *178*, 355-361.
33. Verkman, A.S. *Am. J. Physiol.* 1990, *259*, C375-C388.
34. Lakowicz, J.R.; Hogen, D. *Chem. Phys. Lipids* 1980, *26*, 1-40.
35. Lehrer, S.S. *Biochemistry* 1971, *10*, 3254-3263.
36. Winkler, M.H. *Biochemistry* 1969, *8*, 2586-2590.
37. Harrochs, A.R.; Kearvell, A.; Tickle, K.; Wilkinson, F. *Trans. Faraday Soc.* 1966, *62*, 3393-3399.
38. Eftink, M.R.; Ghiron, C.A. *J. Phys. Chem.* 1976, *80*, 486-493.

39. Eftink, M.R.; Ghiron, C.A. *Biochemistry* 1984, *23*, 3891-3899.
40. Wolfbeis, O.S.; Sharma, A. *Anal. Chim. Acta* 1988, *208*, 53-58.
41. Wolfbeis, O.S.; Posh, H.E.; Kroneis, H.K. *Anal. Chem.* 1985, *57*, 2556-2561.
42. Hitchman, M.L. *Measurement of Dissolved Oxygen*; John Wiley & Sons: New York, NY, 1978.
43. Zhujun, Z.; Seitz, W.R. *Anal. Chem.* 1986, *58*, 220-222.
44. Freeman, T.M.; Seitz, W.R. *Anal. Chem.* 1981, *53*, 98-102.
45. Vanderkooi, J.M.; Maniara, G.; Green, T.J.; Wilson, D.F. *J. Biol. Chem.* 1987, *262*, 5476-5482.
46. Turro, N.J.; Cox, G.S.; Li, X. *Photochem. Photobiol.* 1983, *37*, 149-153.
47. Lee, E.D.; Werner, T.C.; Seitz, W.R. *Anal. Chem.* 1987, *59*, 279-283.
48. Peterson, J.I.; Fitzgerald, R.V.; Buckhold, D.K. *Anal. Chem.* 1984, *56*, 62-67.
49. Gehrich, J.L.; Lubbers, D.W.; Opitz, N.; Hansman, D.R.; Miller, W.W.; Tusa, J.K.; Yafuso, M. *IEEE Trans. Biomed. Eng.* 1986, *33*, 117-132.
50. Opitz, N.; Lubbers, D.W. *Adv. Exp. Med. Biol.* 1987, *215*, 45-50.
51. Lippitsch, M.E.; Pasterhofer, J.; Leiner, M.J.P.; Wolfbeis, O.S. *Anal. Chim. Acta* 1988, *205*, 1-6.
52. Barnikol, W.K.R.; Gaertner, Th.; Weiler, N. *Rev. Sci. Instrum.* 1988, *59(7)*, 1204-1208.
53. Trettnak, W. In *Fluorescence Spectroscopy. New Methods and Applications*; Wolfbeis, O.S., Ed.; Springer-Verlag: Berlin, Heidelberg, 1993, pp. 79-89.
54. Pfeil, A. *J. Am. Chem. Soc.* 1971, *93*, 5395-5397.
55. Demas, J.N.; Diemente, D.; Harris, E.W. *J. Am. Chem. Soc.* 1973, *95*, 6864-6865.
56. Carraway, E.R.; Demas, J.N.; DeGraff, B.A.; Bacon, J.R. *Anal. Chem.* 1991, *63*, 337-342.
57. Demas, J.N.; Harris, E.W.; McBride, R.P. *J. Am. Chem. Soc.* 1977, *99*, 3547-3551.
58. Van Houten, J.; Watts, R.J. *J. Am. Chem. Soc.* 1976, *98*, 4853-4858.
59. Förster, T. *Ann. Phys. (Leipzig)* 1948, *2*, 55-75 (Translated by R.S. Knox).
60. Stryer, L. *Ann. Rev. Biochem.* 1978, *47*, 819-846.
61. Jordan, D.M.; Walt, D.R.; Milanovich, F.P. *Anal. Chem.* 1987, *59*, 437-439.
62. Meadows, D.; Shultz, J.S. *Talanta* 1988, *35*, 145-150.
63. Roe, J.N.; Szoka, F.C.; Verkman, A.S. *Analyst* 1990, *115*, 353-358.
64. Siegmund, H-U.; Becker, A.; Ohst, H.; Sommer, K. *Thin Solid Films* 1992, *210/211*, 480-483.
65. Lakowicz, J.R.; Maliwal, B. *Anal. Chim. Acta* 1993, *271*, 155-164.
66. Lakowicz, J.R.; Szmacinski, H.; Karakelle, M. *Anal. Chim. Acta* 1993, *272*, 179-186.
67. Parker, F.S. *Applications of Infrared, Raman, and Resonance Raman Spectroscopy in Biochemistry*; Plenum Press: New York, NY, 1983; pp 497-505.
68. Lubbers, D.W.; Opitz, N. *Z. Naturforsch.* 1975, *30c*, 532-533.
69. Vurek, G.G.; Feustel, P.J.; Severinghaus, J.W. *Ann. Biomed. Eng.* 1983, *11*, 499-510.

70. Opitz, N.; Lubbers, D.W. *Adv. Exp. Med. Biol.* 1984, *180*, 757-762.
71. Zhujun, Z.; Seitz, W.R. *Anal. Chim. Acta* 1984, *160*, 305-309.
72. Mulkholm, C.; Walt, D.R.; Milanovich, F.P. *Talanta* 1988, *35(2)*, 109-112.
72. Kawabata, Y.; Kamichika, T.; Imasaka, T.; Ishibashi, N. *Anal. Chim. Acta* 1989, *219*, 223-229.
73. Orellana, G.; Moreno-Bondi, M.C.; Segovia, E.; Marazuela, M.D. *Anal. Chem.* 1992, *64*, 2210-2215.
75. Yuan, P.; Walt, D.R. *Anal. Chem.* 1987, *59*, 2391-2394.
76. Jones, T.P.; Porter, M.D. *Anal. Chem.* 1988, *60*, 404-406.
77. Moreno-Bondi, M.C.; Wolfbeis, O.S.; Leiner, M.J.P.; Schaffar, B.P.H. *Anal. Chem.* 1990, *62*, 2377-2380.
78. Trettnak, W.; Leiner, M.J.P.; Wolfbeis, O.S. *Analyst* 1988, *113*, 1519-1523.
79. Trettnak, W.; Leiner, M.J.P.; Wolfbeis, O.S. *Biosensors* 1988, *4*, 15-26.
80. Shichiri, M.; Kawamori, R.; Yamasaki, Y. *Methods in Enzymol.* 1988, *137*, 326-334.
81. Trettnak, W.; Wolfbeis, O.S. *Anal. Chim. Acta* 1989, *221*, 195-203.
82. Bauer, B.; Floyd, T.A. *Anal. Chim. Acta* 1987, *197*, 295-301.
83. Kaiser, N. *J. Horm. Matabol. Res. Suppl.* 1977, *8*, 30-33.
84. Schultz, J.S.; Mansouri, S.; Goldstein, I.J. *Diabetes Care* 1982, *5(3)*, 245-253.
85. Lakowicz, J.R.; Berndt, K.W. *Chem. Phys. Lett.* 1990, *166(3)*, 246-252.
86. Whitaker, J.E.; Haugland, R.P.; Prendergast, G. *Anal. Biochem.* 1991, *194*, 330-334.
87. Miyoshi, N.; Hara, K.; Kimura, S.; Nakanishi, K; Fukuda, M. *Photochem. Photobiol.* 1991, *53*, 415-418.
88. Lakowicz, J.R.; Szmacinski, H.; Nowaczyk, K.; Johnson, M.L. *Cell Calcium* 1992, *13*, 131-147.
89. Lakowicz, J.R; Szmacinski, H.; Johnson, M.L. *J. Fluorescence* 1992, *2*, 47-62.
90. Hirshfield, K.M.; Packard, B.S.; Brand, L. *Biophys. J.* 1992, *61*, A314.
91. Hirshfield, K.M.; Toptygin, D.; Packard, B.S.; Brand, L. *Anal. Biochem.* 1993, *209*, 209-218.
92. Szmacinski, H.; Lakowicz, J.R. *Biophys. J.* 1993, *64*, A108.
93. Koler, E.; Wolfbeis, O.S. In *Fiber Optic Chemical Sensors and Biosensors*; Wolfbeis, O.S., Ed.; CRC Press: Boca Raton, Fl, 1991; Vol. I, pp 306-320.
94. Rothenberg, P.; Glaser, L.; Schlesinger, O.; Cassel, D. *J. Biol. Chem.* 1983, *258*, 12644-12653.
95. Chaillet, J.R.; Boron, W.F. *J. Gen. Physiol.* 1985, *86*, 765-794.
96. Bright, G.R.; Fisher, G.W.; Rogowska, J.; Taylor, D.L. *J. Cell Biol.* 1987, *104*, 1019-1033.
97. Fuh, M-R.S.; Burgess, L-W.; Hirschfeld, T.; Christian, G.D.; Wang, F. *Analyst* 1987, *112*, 1159-1161.
98. MacCraith, B.D.; Ruddy, V.; Potter, C.; McGilp, J.F.; O'Kelly, B. *Proceedings SPIE* 1991, Vol. 1510 (Chemical and Medical Sensors), pp. 104-109.
99. Zhujun, Z.; Seitz, W.R. *Anal. Chim. Acta* 1984, *160*, 47-55.
100. Offenbacher, H.; Wolfbeis, O.S.; Furlinger, E. *Sensors and Actuators* 1986, *9*, 73-84.

101. Tsien R.Y. In *Methods in Cell Biology*; Taylor, D.L.; Wang, Y.-L., Eds.; Academic Press: London, 1989, Vol. 30, pp. 127-156.
102. Thompson, R.B.; Lakowicz, J.R. *Anal. Chem.* 1993, *65*, 853-856.
103. Cobbold, P.H.; Rink, J. *Biochem. J.* 1987, *248*, 313-328.
104. Tsien, R.Y. *Biochemistry* 1980, *19*, 2396-2404.
105. Tsien, R.Y.; Pozzan, T.; Rink, T.J. *J. Cell Biol.* 1982, *94*, 325-334.
106. Grynkiewicz, G.; Poenie, M.; Tsien, R.Y. *J. Biol. Chem.* 1985, *260*, 3440-3450.
107. Minta, A.; Kao, J.P.Y.; Tsien, R.Y. *J. Biol. Chem.* 1989, *264*, 8171-8176.
108. Eberhard, M.; Erne, P. *Biochem. Biophys. Res. Comm.* 1991, *180*, 209-215.
109. Keating, S.M.; Wensel, T.G. *Biophys. J.* 1991, *59*, 186-202.
110. Pinsky, B.G.; Ladasky, J.J.; Lakowicz, J.R.; Berndt, K.; Hoffman, R.A. *Cytometry* 1993, *14*, 123-135.
111. Lakowicz, J.R.; Szmacinski, H.; Nowaczyk, K.; Berndt, K.W.; Johnson, M.L. *Anal. Biochem.* 1992, *202*, 316-330.
112. Lakowicz, J.R.; Szmacinski, H,; Nowaczyk, K.; Johnson, M.L. *Proc. Natl. Acad. Sci. USA* 1992, *89*, 1271-1275.
113. Lakowicz, J.R.; Szmacinski, H, Berndt, K.W. In *Fiber Optic Medical and Fluorescent Sensors and Applications*, SPIE Proceedings, 1992, Vol. 1648, pp.150-163.

RECEIVED May 11, 1993

INDEXES

Author Index

Affiliation Index

Subject Index

Production: Meg Marshall
Indexing: Deborah H. Steiner
Acquisition: Anne Wilson
Cover design: Amy Hayes

Printed and bound by Maple Press, York, PA

Highlights from ACS Books

Good Laboratory Practice Standards: Applications for Field and Laboratory Studies
Edited by Willa Y. Garner, Maureen S. Barge, and James P. Ussary
ACS Professional Reference Book; 572 pp; clothbound ISBN 0–8412–2192–8

Silent Spring Revisited
Edited by Gino J. Marco, Robert M. Hollingworth, and William Durham
214 pp; clothbound ISBN 0–8412–0980–4; paperback ISBN 0–8412–0981–2

The Microkinetics of Heterogeneous Catalysis
By James A. Dumesic, Dale F. Rudd, Luis M. Aparicio, James E. Rekoske,
and Andrés A. Treviño
ACS Professional Reference Book; 316 pp; clothbound ISBN 0–8412–2214–2

Helping Your Child Learn Science
By Nancy Paulu with Margery Martin; Illustrated by Margaret Scott
58 pp; paperback ISBN 0–8412–2626–1

Handbook of Chemical Property Estimation Methods
By Warren J. Lyman, William F. Reehl, and David H. Rosenblatt
960 pp; clothbound ISBN 0–8412–1761–0

Understanding Chemical Patents: A Guide for the Inventor
By John T. Maynard and Howard M. Peters
184 pp; clothbound ISBN 0–8412–1997–4; paperback ISBN 0–8412–1998–2

Spectroscopy of Polymers
By Jack L. Koenig
ACS Professional Reference Book; 328 pp;
clothbound ISBN 0–8412–1904–4; paperback ISBN 0–8412–1924–9

Harnessing Biotechnology for the 21st Century
Edited by Michael R. Ladisch and Arindam Bose
Conference Proceedings Series; 612 pp;
clothbound ISBN 0–8412–2477–3

From Caveman to Chemist: Circumstances and Achievements
By Hugh W. Salzberg
300 pp; clothbound ISBN 0–8412–1786–6; paperback ISBN 0–8412–1787–4

The Green Flame: Surviving Government Secrecy
By Andrew Dequasie
300 pp; clothbound ISBN 0–8412–1857–9

For further information and a free catalog of ACS books, contact:
American Chemical Society
Distribution Office, Department 225
1155 16th Street, NW, Washington, DC 20036
Telephone 800–227–5558

Bestsellers from ACS Books

The ACS Style Guide: A Manual for Authors and Editors
Edited by Janet S. Dodd
264 pp; clothbound ISBN 0–8412–0917–0; paperback ISBN 0–8412–0943–X

The Basics of Technical Communicating
By B. Edward Cain
ACS Professional Reference Book; 198 pp;
clothbound ISBN 0–8412–1451–4; paperback ISBN 0–8412–1452–2

Chemical Activities (student and teacher editions)
By Christie L. Borgford and Lee R. Summerlin
330 pp; spiralbound ISBN 0–8412–1417–4; teacher ed. ISBN 0–8412–1416–6

Chemical Demonstrations: A Sourcebook for Teachers,
Volumes 1 and 2, Second Edition
Volume 1 by Lee R. Summerlin and James L. Ealy, Jr.;
Vol. 1, 198 pp; spiralbound ISBN 0–8412–1481–6;
Volume 2 by Lee R. Summerlin, Christie L. Borgford, and Julie B. Ealy
Vol. 2, 234 pp; spiralbound ISBN 0–8412–1535–9

Chemistry and Crime: From Sherlock Holmes to Today's Courtroom
Edited by Samuel M. Gerber
135 pp; clothbound ISBN 0–8412–0784–4; paperback ISBN 0–8412–0785–2

Writing the Laboratory Notebook
By Howard M. Kanare
145 pp; clothbound ISBN 0–8412–0906–5; paperback ISBN 0–8412–0933–2

Developing a Chemical Hygiene Plan
By Jay A. Young, Warren K. Kingsley, and George H. Wahl, Jr.
paperback ISBN 0–8412–1876–5

Introduction to Microwave Sample Preparation: Theory and Practice
Edited by H. M. Kingston and Lois B. Jassie
263 pp; clothbound ISBN 0–8412–1450–6

Principles of Environmental Sampling
Edited by Lawrence H. Keith
ACS Professional Reference Book; 458 pp;
clothbound ISBN 0–8412–1173–6; paperback ISBN 0–8412–1437–9

Biotechnology and Materials Science: Chemistry for the Future
Edited by Mary L. Good (Jacqueline K. Barton, Associate Editor)
135 pp; clothbound ISBN 0–8412–1472–7; paperback ISBN 0–8412–1473–5

For further information and a free catalog of ACS books, contact:
American Chemical Society
Distribution Office, Department 225
1155 16th Street, NW, Washington, DC 20036
Telephone 800–227–5558